Late Night Thoughts on Listening to Mahler's Ninth Symphony

Also by Lewis Thomas

The Lives of a Cell

The Medusa and the Snail

The Youngest Science

Late Night Thoughts on Listening to Mahler's Ninth Symphony

Lewis Thomas

The Viking Press
New York

To Beryl, again

First published in 1983 by The Viking Press
40 West 23rd Street, New York, N. Y. 10010
Published simultaneously in Canada by
Penguin Books Canada Limited

"Alchemy," "Altruism," "Basic Science and the Pentagon," "Clever Animals," "Falsity and Failure," "Late Night Thoughts on Listening to Mahler's Ninth Symphony," "Making Science Work," "On Medicine and the Bomb," "My Magical Metronome," "On Matters of Doubt," "On the Need for Asylums," "Science and 'Science,'" "Some Scientific Advice," "On Speaking of Speaking," "The Artificial Heart," "The Attic of the Brain," "The Corner of the Eye," "The Lie Detector," and "The Problem of Dementia" appeared originally in Discover; *"On Smell" in the* New England Journal of Medicine *and "The Unforgettable Fire" ("Unacceptable Damage") in* The New York Review of Books.

Grateful acknowledgment is made to Alfred A. Knopf, Inc., for permission to reprint an excerpt from "The Man with the Blue Guitar," from the Collected Poems of Wallace Stevens, *Copyright 1936 by Wallace Stevens and renewed 1964 by Holly Stevens.*

LIBRARY OF CONGRESS CATALOGING IN PUBLICATION DATA
Thomas, Lewis, 1913–
Late night thoughts on listening to Mahler's Ninth Symphony.
1. Science—Philosopy. 2. Biology—Philosophy. I. Title.
Q175.T495 1983 081 83-47932
ISBN 0-670-70390-7

Printed in the United States of America
Set in Linotron Goudy

Contents

Late Night Thoughts on Listening to Mahler's Ninth Symphony

The Unforgettable Fire

The hardest of all tasks for the military people who are occupationally obliged to make plans for wars still to come must be to keep a comprehensive, up-to-date list of guesses as to what the other side might, in one circumstance or another, do. Prudence requires that all sorts of possibilities be kept in mind, including, above all, the "worst-case." In warfare, in this century, the record has already proved that the worst-case will turn out in the end to be the one that happens and, often enough, the one that hadn't been planned for. At the outset of World War I, the British didn't have in mind the outright loss of an entire generation of their best youth, nor did any of the Europeans count on such an unhinging of German society as would lead straight to Hitler. In World War II, when things were being readied, nobody forecast Dresden or Coventry as eventualities to be looked out for and planned against. In

Vietnam, defeat at the end was not anywhere on the United States' list of possible outcomes, nor was what happened later in Cambodia and Laos part of the scenario.

We live today in a world densely populated by human beings living in close communication with one another all over the surface of the planet. Viewed from a certain distance it has the look of a single society, a community, the swarming of an intensely social species trying to figure out ways to become successfully interdependent. We obviously need, at this stage, to begin the construction of some sort of world civilization. The final worst-case for all of us has now become the destruction, by ourselves, of our species.

This will not be a novel event for the planet, if it does occur. The fossil record abounds with sad tales of creatures that must have seemed stunning successes in their heyday, wiped out in one catastrophe after another. The trilobites are everywhere, elegant fossil shells, but nowhere alive. The dinosaurs came, conquered, and then all at once went.

Epidemic disease, meteorite collisions, volcanoes, atmospheric shifts in the levels of carbon dioxide, earthquakes, excessive warming or chilling of the earth's surface are all on the worst-case list for parts of the biosphere, one time or another, but it is unlikely that these can ever be lethal threats to a species as intelligent and resourceful as ours. We will not be wiped off the face of the earth by hard times, no matter how hard; we are tough and resilient animals, good at hard times. If we are to be done in, we will do it ourselves by warfare with thermonuclear weaponry, and it will happen because the military planners, and the governments who pay close attention to them, are guessing at the wrong worst-case. At the moment there are really only two groups, the Russians and us, but soon there will be others, already lining up.

Each side is guessing that the other side will, sooner or later, fire first. To guard against this, each side is hell-bent on achieving a weapons technology capable of two objectives: to prevent the other from firing first by having enough missiles to destroy the first-strike salvo before it is launched (which means, of course, its own first strike) and, as a backup, to have for retaliation a powerful enough reserve to inflict what is called "unacceptable damage" on the other side's people. In today's urban world, this means the cities. The policy revision designated as Presidential Directive 59, issued by the Carter White House in August 1980, stipulates that enemy command and control networks and military bases would become the primary targets in a "prolonged, limited" nuclear war. Even so, *some* cities and towns would inevitably be blown away, then doubtless more, then perhaps all.

The term "unacceptable" carries the implication that there is an acceptable degree of damage from thermonuclear bombs. This suggests that we are moving into an era when the limited use of this kind of weaponry is no longer on the worst-case lists. Strategic weapons are those designed to destroy the whole enemy—armies, navies, cities, and all. Tactical nuclear bombs are something else again, smaller and neater, capable of taking out a fortified point, selectively and delicately removing, say, a tank division. Damage to one's country from strategic weapons may be unacceptable, in these terms, but tactical weapons do not raise this issue.

So it goes. The worst-case is clouds of missiles coming

over the horizon, aimed at the cities. And another sort of merely bad-case might be to *neglect* the advantage of small-scale, surgically precise, tactical weapons, needed at crucial moments on conventional battlefields when things are going against one's side. So it seems.

Perhaps it really is the other way round. The worst of all possible scenarios might be the tactical use of a miniaturized thermonuclear bomb, a mere puff alongside the gigantic things stowed on MIRVs.

When we speak of mere puffs, it is useful to give a backward thought to Hiroshima and Nagasaki. We have gotten used to the notion that the two bombs dropped out of our B-29 bombers on August 6 and 9, 1945, were only primitive precursors of what we have at hand today, relatively feeble instruments, even rather quaint technological antiques, like Tiffany lamps. They were indeed nothing but puffs compared to what we now possess. If we and the Russians were to let everything fly at once, we could do, in a matter of minutes, a million times more damage than was done on those two August mornings long ago.

How do you figure a million times? You have to know, to begin with and in some detail, precisely what the Hiroshima and Nagasaki damage was like, and then try to imagine it a millionfold magnified. You don't even have to do that, if your imagination doesn't stretch that far. Any single one of today's best hydrogen bombs will produce at least one thousand times the lethal blast, heat, and radiation that resulted from the Hiroshima or Nagasaki bombs. Nothing would remain alive, no matter how shielded or "hardened," within an area twelve miles in diameter. Taking just one city—Boston, for example—you can begin to guess with some accuracy what just one modern bomb can do, say, tomorrow morning.

One thing for sure, such a bomb would not leave alive anyone to join a committee to prepare a book of sketches and paintings like those published several years ago in *Unforgettable Fire*. The people whose memories are contained in this book were residents of Hiroshima, most of them somewhere within a radius of two miles or so from the hypocenter, the Aioi bridge in the center of town. They survived, and made their drawings thirty years later. With one of today's bombs they would all have been vaporized within a fraction of a second after the explosion. What they recall most vividly, and draw most heartrendingly, are the deaths all around them, the collapsed buildings, and above all the long black strips of skin hanging from the arms and torsos of those still alive. They remember the utter hopelessness, the inability of anyone to help anyone else, the loneliness of the injured alongside the dying. Reading their accounts and wincing at the pictures, one gains the sure sense that no society, no matter how intricately structured, could have coped with that event. No matter how many doctors and hospitals might have been in place and ready to help with medical technology beforehand (as, for instance, in Boston or Baltimore today), at the moment of the fireball all of that help would have vanished in the new sun. As for the radioactivity, a single case of near-lethal radiation can occasionally be saved today by the full resources of a highly specialized, tertiary hospital unit, with endless transfusions and bone-marrow transplants. But what to do about a thousand such cases all at once, or a hundred thousand? Not to mention the more conventionally maimed and burned people, in the millions.

Words like "disaster" and "catastrophe" are too frivolous for the events that would inevitably follow a war with thermonuclear weapons. "Damage" is not the real term; the language has no word for it. Individuals might survive, but "survival" is itself the wrong word. As to the thought processes of the people in high perches of government who believe that they can hide themselves underground somewhere (they probably can) and emerge later on to take over again the running of society (they cannot, in the death of society) or, more ludicrous, the corporate executives who plan to come deranged out of their underground headquarters already installed in the mountains to reorganize the telephone lines or see to the oil business—these people cannot have thought at all.

Hiroshima and Nagasaki: The Physical, Medical and Social Effects of the Atomic Bombings is a 700-page, flatly written description of what happened in August 1945, containing a few photographs and a great many charts and tables to illustrate the abundant details in the text. Here and there, but in only a few paragraphs of the unemotional, factual text, are sentences that reveal the profundity of revulsion and disgust for this weapon and its use by the United States that still remain in the Japanese mind. It is briefly noted that Hiroshima had been spared the extensive fire-bombing to which most other Japanese cities were being subjected in the 1945 summer—there was an eagerly believed rumor that the Americans were sparing the city out of respect because it was known to be a Buddhist religious center—and then later, soon after the A-bombing, it was realized that the city had been preserved free of conventional damage in order to measure with exactitude the effects of the new bomb. It is tersely recalled that the first American journalists to arrive on the scene, a month after the bombing, were interested only in the extent of physical damage and the evidence of the instrument's great power, and it is further noted that no news about the injuries to the people, especially news about radiation sickness, was allowed by the Allied Occupation. "*On 6 September, 1945, the General Headquarters of the Occupational Forces issued a statement that made it clear that people likely to die from A-bomb afflictions should be left to die. The official attitude . . . was that people suffering from radiation injuries were not worth saving.*" The Japanese nation is of course now a friendly ally of the United States, peacefully linked to this country, but the people remember. The continuing bitterness of that memory runs far deeper than most Americans might guess. Apart from the 370,000 Hiroshima and Nagasaki survivors still alive, whose lasting evidences of physical and psychological damage are exhaustively documented in *Hiroshima and Nagasaki*, the Japanese people at large are appalled that other nations can still be so blind to the horror.

To get back to cases, worst-cases, what would you guess is the worst of all possibilities today, with the United States and the Soviet Union investing every spare dollar and ruble to build new and more powerful armaments, missiles enough to create artificial suns in every habitable place of both countries, and with France, Britain, India, China, South Africa, maybe Israel, and who knows what other country either stockpiling bombs of their own or preparing to build them? Of all the mistakes to be made, which is the worst?

A very bad one, although maybe not the worst of all, is a technical complication not much talked about in public but hanging over all the military scientists like a great net poised to fall at any moment. It is still a theoretical complication, not yet tested, or, for that matter, even testable, but very terrifying indeed. The notion is this: a good-sized nuclear bomb, say, ten megatons, exploded at a very high altitude, 250 miles or so over a country, or a set of several such bombs over a continent, might elicit such a surge of electromagnetic energy in the underlying atmosphere that all electronic devices on the earth below would be put out of commission—or destroyed outright—all computers, radios, telephones, television, all electric grids, all communications beyond the reach of a human shout. None of the buttons pressed in Moscow or Washington, if either lay beneath the rays, would function. The silos would not open on command, or fire their missiles. During this period the affected country would be, in effect, anaesthetized, and the follow-on missiles from the other side could pick off their targets like fruit from a tree. Only the submarine forces, roaming far at sea, would be able to fire back, and their only signal to fire would have to be the total absence of any signal from home. The fate of the aggressor's own cities would then lie at the fingertips of individual submarine commanders, out of touch with the rest of the world, forced to read the meaning of silence.

If the hypothesis is valid, it introduces a new piece of logic into the game. Metal shielding can be used to protect parts of the military communication lines, and fiber optics lines are already replacing some parts. Nuclear-power plants can be partially shielded, perhaps enough to prevent meltdowns everywhere. But no one can yet be certain of protection against this strange new threat.

Any first strike might have to involve this technical maneuver beforehand, with the risk of counterblows from somewhere in the oceans or from surviving intelligence in land-based missiles, but always with the tempting prospect of an enemy country sitting paralyzed like a rabbit in the headlights of a truck. Guessing wrong either way could be catastrophic, and I imagine the War College faculties on both sides are turning the matter over and over in their minds and on their computers, looking for new doctrine.

But all in all, looking ahead, it seems to me that the greatest danger lies in the easy assumption by each government that the people in charge of military policy in any adversary government are not genuine human beings. We make this assumption about the Russians all the time, and I have no doubt they hold the same belief about us. We know ourselves, of course, and take ourselves on faith: Who among us would think of sending off a cluster of missiles to do a million times more damage to a foreign country than was done at Hiroshima, for any reason? None of us, we would all affirm (some of us I fear with fingers crossed). But there are those people on the other side who do not think as we do, we think.

It may be that the road to the end is already being paved, right now, by those tactical "theater" bombs, the little ones, as small and precise as we can make them. If the other side's tanks are gaining on ours, and we are about to lose an action, let them have one! And when they send one of theirs back in retaliation, slightly larger, let them have another, bigger one. Drive them back, we will say, let them have it. A few such exchanges, and off will go the ICBMs, and down will go, limb by limb, all of mankind.

I hope these two books are widely translated, and then propped under the thoughtful, calculating, and expressionless eyes of all the officials in the highest reaches of all governments. They might then begin to think harder than they now think about the future, their own personal future, and about whether a one-time exchange of bombs between countries would leave any government still governing, or any army officer still in command of anything.

Maybe the military people should sit down together on neutral ground, free of politicians and diplomats, perhaps accompanied by their chief medical officers and hospital administrators, and talk together about the matter. They are, to be sure, a strange and unfamiliar lot, unworldly in a certain sense, but they know one another or could at least learn to know one another. After a few days of discussion, unaffectionately and coldly but still linked in a common and ancient professional brotherhood, they might reach the conclusion that the world is on the wrong track, that human beings cannot fight with such weapons and remain human, and that since organized societies are essential for the survival of the profession of arms it is time to stop. It is the generals themselves who should have sense enough to demand a freeze on the development of nuclear arms, and then a gradual, orderly, meticulously scrutinized reduction of such arms. Otherwise, they might as well begin to learn how spears are made, although their chances of living to use them are very thin, not much better than the odds for the rest of us.

Meanwhile, the preparations go on, the dreamlike rituals are rehearsed, and the whole earth is being set up as an altar for a burnt offering, a monstrous human sacrifice to an imagined god with averted eyes. Carved in the stone of the cenotaph in Hiroshima are the words: REST IN PEACE, FOR THE MISTAKE WILL NOT BE REPEATED. The inscription has a life of its own. Intended first as a local prayer and promise, it has already changed its meaning into a warning, and is now turning into a threat.

The Corner of the Eye

There are some things that human beings can see only out of the corner of the eye. The niftiest examples of this gift, familiar to all children, are small, faint stars. When you look straight at one such star, it vanishes; when you move your eyes to stare into the space nearby, it reappears. If you pick two faint stars, side by side, and focus on one of the pair, it disappears and now you can see the other in the corner of your eye, and you can move your eyes back and forth, turning off the star in the center of your retina and switching the other one on. There is a physiological explanation for the phenomenon: we have more rods, the cells we use for light perception, at the periphery of our retinas, more cones, for perceiving color, at the center.

Something like this happens in music. You cannot really hear certain sequences of notes in a Bach fugue unless at the same time there are other notes being sounded, dominating the field. The real meaning in music comes from tones only audible in the corner of the mind.

I used to worry that computers would become so powerful and sophisticated as to take the place of human minds. The notion of Artificial Intelligence used to scare me half to death. Already, a large enough machine can do all sorts of intelligent things beyond our capacities: calculate in a split second the answers to mathematical problems requiring years for a human brain, draw accurate pictures from memory, even manufacture successions of sounds with a disarming resemblance to real music. Computers can translate textbooks, write dissertations of their own for doctorates, even speak in machine-tooled, inhuman phonemes any words read off from a printed page. They can communicate with one another, holding consultations and committee meetings of their own in networks around the earth.

Computers can make errors, of course, and do so all the time in small, irritating ways, but the mistakes can be fixed and nearly always are. In this respect they are fundamentally inhuman, and here is the relaxing thought: computers will not take over the world, they cannot replace us, because they are not designed, as we are, for ambiguity.

Imagine the predicament faced by a computer programmed to make language, not the interesting communication in sounds made by vervets or in symbols by brilliant chimpanzee prodigies, but real human talk. The grammar would not be too difficult, and there would be no problem in constructing a vocabulary of etymons, the original, pure, unambiguous words used to name real things. The impossibility would come in making the necessary mistakes we humans make with words instinctively, intuitively, as we build our kinds of language, changing the meanings to imply quite different things, constructing and elaborating the varieties of ambiguity without which speech can never become human speech.

Look at the record of language if you want to glimpse the special qualities of the human mind that lie beyond the reach of any machine. Take, for example, the metaphors we use in everyday speech to tell ourselves who we are, where we live, and where we come from.

The earth is a good place to begin. The word "earth" is used to name the ground we walk on, the soil in which we grow plants or dig clams, and the planet itself; we also use it to describe all of humanity ("the whole earth responds to the beauty of a child," we say to each other).

The earliest word for earth in our language was the Indo-European root *dhghem*, and look what we did with it. We turned it, by adding suffixes, into *humus* in Latin; today we call the complex polymers that hold fertile soil together "humic" acids, and somehow or other the same root became "humility." With another suffix the word became "human." Did the earth become human, or did the human emerge from the earth? One answer may lie in that nice cognate word "humble." "Humane" was built on, extending the meaning of both the earth and ourselves. In ancient Hebrew, *adamha* was the word for earth, *adam* for man. What computer could run itself through such manipulations as those?

We came at the same system of defining ourselves from the other direction. The word *wiros* was the first root for man; it took us in our vanity on to "virile" and "virtue," but also turned itself into the Germanic word *weraldh*, meaning the life of man, and thence in English to our word "world."

There is a deep hunch in this kind of etymology. The world of man derives from this planet, shares origin with the life of the soil, lives in humility with all the rest of life. I cannot imagine programming a computer to think up an idea like that, not a twentieth-century computer, anyway.

The world began with what it is now the fashion to call the "Big Bang." Characteristically, we have assigned the wrong words for the very beginning of the earth and ourselves, in order to evade another term that would cause this century embarrassment. It could not, of course, have been a bang of any sort, with no atmosphere to conduct the waves of sound, and no ears. It was something else, occurring in the most absolute silence we can imagine. It was the Great Light.

We say it had been chaos before, but it was not the kind of place we use the word "chaos" for today, things tumbling over each other and bumping around. Chaos did not have that meaning in Greek; it simply meant empty.

We took it, in our words, from chaos to cosmos, a word that simply meant order, cosmetic. We perceived the order in surprise, and our cosmologists and physicists continue to find new and astonishing aspects of the order. We made up the word "universe" from the whole affair, meaning literally turning everything into one thing. We used to say it was a miracle, and we still permit ourselves to refer to the whole universe as a marvel, holding in our unconscious minds the original root meaning of these two words, miracle and marvel—from the ancient root word *smei*, signifying a smile. It immensely pleases a human being to see something never seen before, even more to learn something never known before, most of all to think something never thought before. The rings of Saturn are the latest surprise. All my physicist friends are enchanted by this phenomenon, marveling at the small violations of the laws of planetary mechanics, shocked by the unaccountable braids and spokes stuck there among the rings like graffiti. It is nice for physicists to see something new and inexplicable; it means that the laws of nature are once again about to be amended by a new footnote.

The greatest surprise of all lies within our own local, suburban solar system. It is not Mars; Mars was surprising in its way but not flabbergasting; it was a disappointment not to find evidences of life, and there was some sadness in the pictures sent back to earth from the Mars Lander, that lonely long-legged apparatus poking about with its jointed arm, picking up sample after sample of the barren Mars soil, looking for any flicker of life and finding none; the only sign of life on Mars was the Lander itself, an extension of the human mind all the way from earth to Mars, totally alone.

Nor is Saturn the great surprise, nor Jupiter, nor Venus, nor Mercury, nor any of the glimpses of the others.

The overwhelming astonishment, the queerest structure we know about so far in the whole universe, the greatest of all cosmological scientific puzzles, confounding all our efforts to comprehend it, is the earth. We are only now beginning to appreciate how strange and splendid it is, how it catches the breath, the loveliest object afloat around the sun, enclosed in its own blue bubble of atmosphere, manufacturing and breathing its own oxygen, fixing its own nitrogen from the air into its own soil, generating its own weather at the surface of its rain forests, constructing its own carapace from living parts: chalk cliffs, coral reefs, old fossils from earlier forms of life now covered by layers of new life meshed together around the globe, Troy upon Troy.

Seen from the right distance, from the corner of the eye of an extraterrestrial visitor, it must surely seem a single creature, clinging to the round warm stone, turning in the sun.

Making Science Work

For about three centuries we have been doing science, trying science out, using science for the construction of what we call modern civilization. Every dispensable item of contemporary technology, from canal locks to dial telephones to penicillin to the Mars Lander, was pieced together from the analysis of data provided by one or another series of scientific experiments—also the technologies we fear the most for the threat they pose to civilization: radioactivity from the stored, stacked bombs or from leaking, flawed power plants, acid rain, pesticides, leached soil, depleted ozone, and increased carbon dioxide in the outer atmosphere.

Three hundred years seems a long time for testing a new approach to human interliving, long enough to settle back for critical appraisal of the scientific method, maybe even long enough to vote on whether to go on with it or not.

There is an argument. Voices have been raised in protest since the beginning, rising in pitch and violence in the nineteenth century during the early stages of the industrial revolution, summoning urgent crowds into the streets any day these days on the issue of nuclear energy. Give it back, say some of the voices, it doesn't really work, we've tried it and it doesn't work, go back three hundred years and start again on something else less chancy for the race of man.

The scientists disagree, of course, partly out of occupational bias, but also from a different way of viewing the course and progress of science in the past fifty years. As they see it, science is just at its beginning. The principal discoveries in this century, taking all in all, are the glimpses of the depth of our ignorance about nature. Things that used to seem clear and rational, matters of absolute certainty—Newtonian mechanics, for example—have slipped through our fingers, and we are left with a new set of gigantic puzzles, cosmic uncertainties, ambiguities; some of the laws of physics are amended every few years, some are canceled outright, some undergo revised versions of legislative intent as if they were acts of Congress.

In biology, it is one stupefaction after another. Just thirty years ago we called it a biological revolution when the fantastic geometry of the DNA molecule was exposed to public view and the linear language of genetics was decoded. For a while things seemed simple and clear; the cell was a neat little machine, a mechanical device ready for taking to pieces and reassembling, like a tiny watch. But just in the last few years it has become almost imponderably complex, filled with strange parts whose functions are beyond today's imagining. DNA is itself no longer a straightforward set of instructions on a tape. There are long strips of what seem nonsense in between the genes, edited out for the assembly of proteins but essential nonetheless for the process of assembly; some genes are called jumping genes, moving from one segment of DNA to another, rearranging the messages, achieving instantly a degree of variability that we once thought would require eons of evolution. The cell membrane is no longer a simple skin for the cell; it is a fluid mosaic, a sea of essential mobile signals, an organ in itself. Cells communicate with one another, exchange messages like bees in a hive, regulate one another. Genes are switched on, switched off, by molecules from the outside whose nature is a mystery; somewhere inside are switches which, when thrown one way or the other, can transform any normal cell into a cancer cell, and sometimes back again.

It is not just that there is more to do, there is everything to do. Biological science, with medicine bobbing somewhere in its wake, is under way, but only just under way. What lies ahead, or what *can* lie ahead if the efforts in basic research are continued, is much more than the conquest of human disease or the amplification of agricultural technology or the cultivation of nutrients in the sea. As we learn more about the fundamental processes of living things in general we will learn more about ourselves, including perhaps the ways in which our brains, unmatched by any other neural structures on the planet, achieve the earth's awareness of itself. It may be too much to say that we will become wise through such endeavors, but we can at least come into possession of a level of information upon which a new kind of wisdom might be based. At the moment we are an ignorant species, flummoxed by the puzzles of who we are, where we came from, and what we are for. It is a gamble to bet on science for moving ahead, but it is, in my view, the only game in town.

The near views in our instruments of the dead soil of Mars, the bizarre rings of Saturn, and the strange surfaces of Saturn, Jupiter, Venus, and the rest, literally unearthly, are only brief glances at what is ahead for mankind in the exploration of our own solar system. In theory, there is no reason why human beings cannot make the same journeys in person, or out beyond into the galaxy.

We will solve our energy problems by the use of science, and in no other way. The sun is there, to be sure, ready for tapping, but we cannot sit back in the lounges of political lobbies and make guesses and wishes; it will take years, probably many years, of research. Meanwhile, there are other possibilities needing deeper exploration. Nuclear fission power, for all its present disadvantages, including where on earth to put the waste, can be made safer and more reliable by better research, while hydrogen fusion, inexhaustibly fueled from the oceans and much safer than fission, lies somewhere ahead. We may learn to produce vast amounts of hydrogen itself, alcohol or methane, when we have learned more about the changeable genes of single-celled microorganisms. If we are to continue to burn coal in large amounts, we will need research models for predicting how much more carbon dioxide we can inject into the planet's atmosphere before we run into the danger of melting the ice shelves of western Antarctica and flooding all our coasts. We will need science to protect us against ourselves.

It has become the fashion to express fear of computers—the machines will do our thinking, quicker and better than human thought, construct and replicate themselves, take over and eventually replace us—that sort of thing. I confess to apprehensions of my own, but I have a hunch that those are on my mind because I do not know enough about computers. Nor, perhaps, does anyone yet, not even the computer scientists themselves. For my comfort, I know for sure only one thing about the computer networks now being meshed together like interconnected ganglia around the earth: what they contain on their microchips are bits of information put there by human minds; perhaps they will do something like thinking on their own, but it will still be a cousin of human thought once removed and, because of newness, potentially of immense usefulness.

The relatively new term "earth science" is itself an encouragement. It is nice to know that our own dear planet has become an object of as much obsessive interest to large bodies of professional researchers as a living cell, and almost as approachable for discovering the details of how it works. Satellites scrutinize it all day and night, recording the patterns of its clouds, the temperatures at all parts of its surface, the distribution and condition of its forests, crops, waterways, cities, and barren places. Seismologists and geologists have already surprised themselves over and over again, probing the movement of crustal plates afloat on something or other, maybe methane, deep below the surface, meditating the evidences now coming in for the reality and continuing of continental drift, and calculating with increasing precision the data that describe the mechanisms involved in earthquakes. Their instruments are becoming as neat and informative as medicine's CAT scanners; the earth has deep secrets still, but they are there for penetrating.

The astronomers have long since become physicists, the physicists are astronomers; both are, as well, what we used to call chemists, examining the levels of ammonia or formaldehyde in clouds drifting billions of light-years away, measuring the concentrations of methane in the nearby atmosphere of Pluto, running into paradoxes. Contemporary physics lives off paradox. Niels Bohr said that a great truth is one for which the opposite is also a great truth. There are not so many neutrinos coming from our sun as there ought to be; something has gone wrong, not with the sun but with our knowledge. There are radioastronomical instruments for listening to the leftover sounds of the creation of the universe; the astronomers are dumbstruck, they can hardly hear themselves think.

The social scientists have a long way to go to catch up, but they may be up to the most important scientific business of all, if and when they finally get down to the right questions. Our behavior toward each other is the strangest, most unpredictable, and almost entirely unaccountable of all the phenomena with which we are obliged to live. In all of nature there is nothing so threatening to humanity as humanity itself. We need, for this most worrying of puzzles, the brightest and youngest of our most agile minds, capable of dreaming up ideas not dreamed before, ready to carry the imagination to great depths and, I should hope, handy with big computers but skeptical about long questionnaires and big numbers.

Fundamental science did not become a national endeavor in this country until the time of World War II, when it was pointed out by some influential and sagacious advisers to the government that whatever we needed for the technology of warfare could be achieved only after the laying of a solid foundation of basic research. During the Eisenhower administration a formal mechanism was created in the White House for the explicit purpose of furnishing scientific advice to the President, the President's Science Advisory Committee (PSAC), chaired by a new administration officer, the Science Adviser. The National Institutes of Health, which had existed before the war as a relatively small set of laboratories for research on cancer and infectious disease, expanded rapidly in the postwar period to encompass all disciplines of biomedical science. The National Science Foundation was organized specifically for the sponsorship of basic science. Each of the federal departments and agencies developed its own research capacity, relevant to its mission; the programs of largest scale were those in defense, agriculture, space, and atomic energy.

Most of the country's basic research has been carried out by the universities, which have as a result become increasingly dependent on the federal government for their sustenance, even their existence, to a degree now causing alarm signals from the whole academic community. The ever-rising costs of doing modern science, especially the prices of today's sophisticated and complex instruments, combined with the federal efforts to reduce all expenditures, are placing the universities in deep trouble. Meanwhile, the philanthropic foundations, which were the principal source of funds for university research before the war, are no longer capable of more than a minor contribution to science.

Besides the government's own national laboratories and the academic institutions there is a third resource for the country's scientific enterprise—industry. Up to very recently, industrial research has been conducted in relative isolation, unconnected with the other two. There are signs that this is beginning to change, and the change should be a source of encouragement for the future. Some of the corporations responsible for high technology, especially those involved in energy, have formed solid linkages with a few research universities—MIT and Cal Tech, for example—and are investing substantial sums in long-range research in physics and chemistry. Several pharmaceutical companies have been investing in fundamental biomedical research in association with medical schools and private research institutions.

There needs to be much more of this kind of partnership. The nation's future may well depend on whether we can set up within the private sector a new system for collaborative research. Although there are some promising partnership ventures now in operation, they are few in number; within industry the tendency remains to concentrate on applied research and development, excluding any consideration of basic science. The academic community tends, for its part, to stay out of fields closely related to the development of new products. Each side maintains adversarial and largely bogus images of the other, money-makers on one side and impractical academics on the other. Meanwhile, our competitors in Europe and Japan have long since found effective ways to link industrial research to government and academic science, and they may be outclassing this country before long. In some fields, most conspicuously the devising and production of new scientific instruments, they have already moved to the front.

There are obvious difficulties in the behavior of the traditional worlds of research in the United States. Corporate research is obliged by its nature to concentrate on profitable products and to maintain a high degree of secrecy during the process; academic science, by its nature, must be carried out in the open and depends for its progress on the free exchange of new information almost at the moment of finding. But these are not impossible barriers to collaboration. Industry already has a life-or-death stake in what will emerge from basic research in the years ahead; there can be no more prudent investment for the corporate world, and the immediate benefit for any corporation in simply having the "first look" at a piece of basic science would be benefit enough in the long run. The university science community, for all the talk of ivory towers, hankers day and night for its work to turn out useful; a close working connection with industrial researchers might well lead to an earlier perception of potential applicability than is now the case.

The age of science did not really begin three hundred years ago. That was simply the time when it was realized that human curiosity about the world represented a deep wish, perhaps embedded somewhere in the chromosomes of human beings, to learn more about nature by experiment and the confirmation of experiment. The doing of science on a scale appropriate to the problems at hand was launched only in the twentieth century and has been moving into high gear only within the last fifty years. We have not lacked explanations at any time in our recorded history, but now we must live and think with the new habit of requiring reproducible observations and solid facts for the explanations. It is not as easy a time for us as it used to be: we are raised through childhood in skepticism and disbelief; we feel the need of proofs all around, even for matters as deep as the working of our own consciousness, where there is as yet no clear prospect of proof about anything. Uncertainty, disillusion, and despair are prices to be paid for living in an age of science. Illumination is the product sought, but it comes in small bits, only from time to time, not ever in broad, bright flashes of public comprehension, and there can be no promise that we will ever emerge from the great depths of the mystery of being.

Nevertheless, we have started to do science on a world scale, and to rely on it, and hope for it. Not just the scientists, everyone, and not for the hope of illumination, but for the sure predictable prospect of new technologies, which have always come along, like spray in the wake of science. We need better ways of predicting how a piece of new technology is likely to turn out, better measures available on an international level to shut off the ones that carry hazard to the life of the planet (including, but perhaps not always so much *first of all*, as is usually the only consideration, our own species' life). We will have to go more warily with technology in the future, for the demands will be increasing and the stakes will be very high. Instead of coping, or trying to cope, with the wants of four billion people, we will very soon be facing the needs, probably desperate, of double that number and, soon thereafter, double again. The real challenge to human ingenuity, and to science, lies in the century to come.

I cannot guess at the things we will need to know from science to get through the time ahead, but I am willing to make one prediction about the method: we will not be able to call the shots in advance. We cannot say to ourselves, we need this or that sort of technology, therefore we should be doing this or that sort of science. It does not work that way. We will have to rely, as we have in the past, on science in general, and on basic, undifferentiated science at that, for the new insights that will open up the new opportunities for technological development. Science is useful, indispensable sometimes, but whenever it moves forward it does so by producing a surprise; you cannot specify the surprise you'd like. Technology should be watched closely, monitored, criticized, even voted in or out by the electorate, but science itself must be given its head if we want it to work.

Alchemy

Alchemy began long ago as an expression of the deepest and oldest of human wishes: to discover that the world makes sense. The working assumption—that everything on earth must be made up from a single, primal sort of matter—led to centuries of hard work aimed at isolating the original stuff and rearranging it to the alchemists' liking. If it could be found, nothing would lie beyond human grasp. The transmutation of base metals to gold was only a modest part of the prospect. If you knew about the fundamental substance, you could do much more than make simple money: you could boil up a cure-all for every disease affecting humankind, you could rid the world of evil, and, while doing this, you could make a universal solvent capable of dissolving anything you might want to dissolve. These were heady ideas, and generations of alchemists worked all their lives trying to reduce matter to its ultimate origin.

To be an alchemist was to be a serious professional, requiring long periods of apprenticeship and a great deal of late-night study. From the earliest years of the profession, there was a lot to read. The documents can be traced back to Arabic, Latin, and Greek scholars of the ancient world, and beyond them to Indian Vedic texts as far back as the tenth century B.C. All the old papers contain a formidable array of information, mostly expressed in incantations, which were required learning for every young alchemist and, by design, incomprehensible to everyone else. The word "gibberish" is thought by some to refer back to Jabir ibn Hayyan, an eighth-century alchemist, who lived in fear of being executed for black magic and worded his doctrines so obscurely that almost no one knew what he was talking about.

Indeed, black magic was what most people thought the alchemists were up to in their laboratories, filled with the fumes of arsenic, mercury, and sulphur and the bubbling infusions of all sorts of obscure plants. We tend to look back at them from today's pinnacle of science as figures of fun, eccentric solitary men wearing comical conical hats, engaged in meaningless explorations down one blind alley after another. It was not necessarily so: the work they were doing was hard and frustrating, but it was the start-up of experimental chemistry and physics. The central idea they were obsessed with—that there is a fundamental, elementary particle out of which everything in the universe is made—continues to obsess today's physicists.

They never succeeded in making gold from base metals,

nor did they find a universal elixir in their plant extracts; they certainly didn't rid the world of evil. What they did accomplish, however, was no small thing: they got the work going. They fiddled around in their laboratories, talked at one another incessantly, set up one crazy experiment after another, wrote endless reams of notes, which were then translated from Arabic to Greek to Latin and back again, and the work got under way. More workers became interested and then involved in the work, and, as has been happening ever since in science, one thing led to another. As time went on and the work progressed, error after error, new and accurate things began to turn up. Hard facts were learned about the behavior of metals and their alloys, the properties of acids, bases, and salts were recognized, the mathematics of thermodynamics were worked out, and, with just a few jumps through the centuries, the helical molecule of DNA was revealed in all its mystery.

The current anxieties over what science may be doing to human society, including the worries about technology, are no new thing. The third-century Roman emperor Diocletian decreed that all manuscripts dealing with alchemy were to be destroyed, on grounds that such enterprises were against nature. The work went on in secrecy, and, although some of the material was lost, a great deal was translated into other languages, passed around, and preserved.

The association of alchemy with black magic has persisted in the public mind throughout the long history of the endeavor, partly because the objective—the transmutation of one sort of substance to another—seemed magical by definition. Partly also because of the hybrid term: *al* was simply the Arabic article, but *chemy* came from a word meaning "the black land," *Khemia*, the Greek name for Egypt. Another, similar-sounding word, *khumeia*, meant an infusion or elixir, and this was incorporated as part of the meaning. The Egyptian origin is very old, extending back to Thoth, the god of magic (who later reappeared as Hermes Trismegistus, master of the hermetic seal required by alchemists for the vacuums they believed were needed in their work). The notion of alchemy may be as old as language, and the idea that language and magic are somehow related is also old. "Grammar," after all, was a word used in the Middle Ages to denote high learning, but it also implied a practicing familiarity with alchemy. *Gramarye*, an older term for grammar, signified occult learning and necromancy. "Glamour," of all words, was the Scottish word for grammar, and it meant, precisely, a spell, casting enchantment.

Medicine, from its dark origins in old shamanism millennia ago, became closely linked in the Middle Ages with alchemy. The preoccupation of alchemists with metals and their properties led to experiments—mostly feckless ones, looking back—with the therapeutic use of all sorts of metals. Paracelsus, a prominent physician of the sixteenth century, achieved fame from his enthusiastic use of mercury and arsenic, based on what now seems a wholly mystical commitment to alchemical philosophy as the key to understanding the universe and the human body simultaneously. Under his influence, three centuries of patients with all varieties of illness were treated with strong potions of metals, chiefly mercury, and vigorous purgation became standard medical practice.

Physics and chemistry have grown to scientific maturity,

medicine is on its way to growing up, and it is hard to find traces anywhere of the earlier fumblings toward a genuine scientific method. Alchemy exists only as a museum piece, an intellectual fossil, so antique that we no longer need be embarrassed by the memory, but the memory is there. Science began by fumbling. It works because the people involved in it work, and *work together*. They become excited and exasperated, they exchange their bits of information at a full shout, and, the most wonderful thing of all, they keep *at* one another.

Something rather like this may be going on now, without realizing it, in the latest and grandest of all fields of science. People in my field, and some of my colleagues in the real "hard" sciences such as physics and chemistry, have a tendency to take lightly and often disparagingly the efforts of workers in the so-called social sciences. We like to refer to their data as soft. We do not acknowledge as we should the differences between the various disciplines within behavioral research—we speak of analytical psychiatry, sociology, linguistics, economics, and computer intelligence as though these inquiries were all of a piece, with all parties wearing the same old comical conical hats. It is of course not so. The principal feature that the social sciences share these days is the attraction they exert on considerable numbers of students, who see the prospect of exploring human behavior as irresistible and hope fervently that a powerful scientific method for doing the exploring can be worked out. All of the matters on the social-science agenda seem more urgent to these young people than they did at any other time in human memory. It may turn out, years hence, that a solid discipline of human science will have come into existence, hard as quantum physics, filled with deep insights, plagued as physics still is by ambiguities but with new rules and new ways of getting things done. Like, for instance, getting rid of thermonuclear weapons, patriotic rhetoric, and nationalism all at once. If anything like this does turn up we will be looking back at today's social scientists, and their close colleagues the humanists, as having launched the new science in a way not all that different from the accomplishment of the old alchemists, by simply working on the problem—this time, the fundamental, primal universality of the human mind.

Clever Animals

Scientists who work on animal behavior are occupationally obliged to live chancier lives than most of their colleagues, always at risk of being fooled by the animals they are studying or, worse, fooling themselves. Whether their experiments involve domesticated laboratory animals or wild creatures in the field, there is no end to the surprises that an animal can think up in the presence of an investigator. Sometimes it seems as if animals are genetically programmed to puzzle human beings, especially psychologists.

The risks are especially high when the scientist is engaged in training the animal to do something or other and must bank his professional reputation on the integrity of his experimental subject. The most famous case in point is that of Clever Hans, the turn-of-the-century German horse now immortalized in the lexicon of behavioral science by the technical term, the "Clever Hans Error." The horse, owned and trained by Herr von Osten, could not only solve complex arithmetical problems, but even read the instructions on a blackboard and tap out infallibly, with one hoof, the right answer. What is more, he could perform the same computations when total strangers posed questions to him, with his trainer nowhere nearby. For several years Clever Hans was studied intensively by groups of puzzled scientists and taken seriously as a horse with something very like a human brain, quite possibly even better than human. But finally in 1911, it was discovered by Professor O. Pfungst that Hans was not really doing arithmetic at all; he was simply observing the behavior of the human experimenter. Subtle, unconscious gestures—nods of the head, the holding of breath, the cessation of nodding when the correct count was reached—were accurately read by the horse as cues to stop tapping.

Whenever I read about that phenomenon, usually recounted as the exposure of a sort of unconscious fraud on the part of either the experimenter or the horse or both, I wish Clever Hans would be given more credit than he generally gets. To be sure, the horse couldn't really do arithmetic, but the record shows that he was considerably better at observing human beings and interpreting their behavior than humans are at comprehending horses or, for that matter, other humans.

Cats are a standing rebuke to behavioral scientists wanting to know how the minds of animals work. The mind of a cat is an inscrutable mystery, beyond human reach, the least human of all creatures and at the same time, as any cat owner will attest, the most intelligent. In 1979, a paper was published in *Science* by B. R. Moore and S. Stuttard entitled "Dr. Guthrie and Felis domesticus or: tripping over the cat," a wonderful account of the kind of scientific mischief native to this species. Thirty-five years ago, E. R. Guthrie and G. P. Horton described an experiment in which cats were placed in a glass-fronted puzzle box and trained to find their way out by jostling a slender vertical rod at the front of the box, thereby causing a door to open. What interested these investigators was not so much that the cats could learn to bump into the vertical rod, but that before doing so each animal performed a long ritual of highly stereotyped movements, rubbing their heads and backs against the front of the box, turning in circles, and finally touching the rod. The experiment has ranked as something of a classic in experimental psychology, even raising in some minds the notion of a ceremony of superstition on the part of cats: before the rod will open the door, it is necessary to go through a magical sequence of motions.

Moore and Stuttard repeated the Guthrie experiment, observed the same complex "learning" behavior, but then discovered that it occurred only when a human being was visible to the cat. If no one was in the room with the box, the cat did nothing but take naps. The sight of a human being was all that was needed to launch the animal on the series of sinuous movements, rod or no rod, door or no door. It was not a learned pattern of behavior, it was a cat greeting a person.

The French investigator R. Chauvin was once engaged in a field study of the boundaries of ant colonies and enlisted the help of some enthusiastic physicists equipped with radioactive compounds and Geiger counters. The ants of one anthill were labeled and then tracked to learn whether they entered the territory of a neighboring hill. In the middle of the work the physicists suddenly began leaping like ballet dancers, terminating the experiment, while hundreds of ants from both colonies swarmed over their shoes and up inside their pants. To Chauvin's ethological eye it looked like purposeful behavior on both sides.

Bees are filled with astonishments, confounding anyone who studies them, producing volumes of anecdotes. A lady of our acquaintance visited her sister, who raised honeybees in northern California. They left their car on a side road, suited up in protective gear, and walked across the fields to have a look at the hives. For reasons unknown, the bees were in a furious mood that afternoon, attacking in platoons, settling on them from all sides. Let us walk away slowly, advised the beekeeper sister, they'll give it up sooner or later. They walked until bee-free, then circled the fields and went back to the car, and found the bees there, waiting for them.

There is a new bee anecdote for everyone to wonder about. It was reported from Brazil that male bees of the plant-pollinating euglossine species are addicted to DDT. Houses that had been sprayed for mosquito control in the Amazonas region were promptly invaded by thousands of bees that gathered on the walls, collected the DDT in pouches on their hind legs, and flew off with it. Most of the houses were virtually stripped of DDT during the summer months, and the residents in the area complained bitterly of the noise. There is as yet no explanation for this behavior. They are not harmed by the substance; while a honeybee is quickly killed by as little as six micrograms of DDT, these bees can cart away two thousand micrograms without being discommoded. Possibly the euglossine bees like the taste of DDT or its smell, or maybe they are determined to protect other insect cousins. Nothing about bees, or other animals, seems beyond imagining.

On Smell

The vacuum cleaner turned on in the apartment's back bedroom emits a high-pitched lament indistinguishable from the steam alarm on the teakettle in the kitchen, and the only way of judging whether to run to the stove is to consult one's watch: there is a time of day for the vacuum cleaner, another time for the teakettle. The telephone in the guest bedroom sounds like the back-door bell, so you wait for the second or third ring before moving. There is a random crunching sound in the vicinity of the front door, resembling an assemblage of people excitedly taking off galoshes, but when listened to carefully it is recognizable as a negligible sound, needing no response, made by the ancient elevator machinery in the wall alongside the door. So it goes. We learn these things from day to day, no trick to it. Sometimes the sounds around our lives become novel confusions, harder to sort out: the family was once given a talking crow named Byron for Christmas, and this animal imitated every nearby sound with such accuracy that the household was kept constantly on the fly, answering doors and telephones, oiling hinges, looking out the window for falling bodies, glancing into empty bathrooms for the sources of flushing.

We are not so easily misled by vision. Most of the things before our eyes are plainly there, not mistakable for other things except for the illusions created for pay by professional magicians and, sometimes, the look of the lights of downtown New York against a sky so black as to make it seem a near view of eternity. Our eyes are not easy to fool.

Smelling is another matter. I should think we might fairly gauge the future of biological science, centuries ahead, by estimating the time it will take to reach a complete, comprehensive understanding of odor. It may not seem a profound enough problem to dominate all the life sciences, but it contains, piece by piece, all the mysteries. Smoke: tobacco burning, coal smoke, wood-fire smoke, leaf smoke. Most of all, leaf smoke. This is the only odor I can *will* back to consciousness just by thinking about it. I can sit in a chair, thinking, and call up clearly to mind the smell of burning autumn leaves, coded and stored away somewhere in a temporal lobe, firing off explosive signals into every part of my right hemisphere. But nothing else: if I try to recall the thick smell of Edinburgh in winter, or the accidental burning of a plastic comb, or a rose, or a glass of wine, I cannot do this; I can get a clear picture of any face I feel like remembering, and I can hear whatever Beethoven quartet I want to recall, but except for the leaf bonfire I cannot really remember a smell in its absence. To be sure, I know the odor of cinnamon or juniper and can name such things with accuracy when they turn up in front of my nose, but I cannot imagine them into existence.

The act of smelling something, anything, is remarkably like the act of thinking itself. Immediately, at the very moment of perception, you can feel the mind going to work, sending the odor around from place to place, setting off complex repertoires throughout the brain, polling one center after another for signs of recognition, old memories, connections. This is as it should be, I suppose, since the cells that do the smelling are themselves proper brain cells, the only neurones whose axones carry information picked up at first hand in the outside world. Instead of dendrites they have cilia, equipped with receptors for all sorts of chemical stimuli, and they are in some respects as mysterious as lymphocytes. There are reasons to believe that each of these neurones has its own specific class of receptors; like lymphocytes, each cell knows in advance what it is looking for; there are responder and nonresponder cells for different classes of odorant. And they arc also the only brain neurones that replicate themselves; the olfactory receptor cells of mice turn over about once every twenty-eight days. There may be room for a modified version of the clonal-selection theory to explain olfactory learning and adaptation. The olfactory receptors of mice can smell the difference between self and nonself, a discriminating gift coded by the same H-2 gene locus governing homograft rejection. One wonders whether lymphocytes in the mucosa may be carrying along this kind of genetic information to donate to new generations of olfactory receptor cells as they emerge from basal cells.

The most medically wonderful of all things about these brain cells is that they do not become infected, not very often anyway, despite their exposure to all the microorganisms in the world of the nose. There must exist, in the mucus secretions bathing this surface of the brain, the most extraordinary antibiotics, including eclectic antiviral substances of some sort.

If you are looking about for things to even out the disparity between the brains of ordinary animals and the great minds of ourselves, the superprimate humans, this apparatus is a good one to reflect on in humility. Compared to the common dog, or any rodent in the field, we are primitive, insensitive creatures, biological failures. Heaven knows how much of the world we are missing.

I suppose if we tried we could improve ourselves. There are, after all, some among our species with special gifts for smelling—perfume makers, tea tasters, whiskey blenders—and it is said that these people can train themselves to higher and higher skills by practicing. Perhaps, instead of spending the resources of our huge cosmetic industry on chemicals for the disguising or outright destruction of odors we should be studying ways to enhance the smell of nature, facing up to the world.

In the meantime, we should be hanging on to some of the few great smells left to us, and I would vote for the preservation of leaf bonfires, by law if necessary. This one is pure pleasure, fetched like music intact out of numberless modular columns of neurones filled chockablock with all the natural details of childhood, firing off memories in every corner of the brain. An autumn curbside bonfire has everything needed for education: danger, surprise (you know in advance that if you poke the right part of the base of leaves with the right kind of stick, a blinding flare of heat and fragrance will follow instantly, but it is still an astonishment when it happens), risk, and victory over odds (if you jump across at precisely the right moment the flare and sparks will miss your pants), and above all the aroma of comradeship (if you smell that odor in the distance you know that there are friends somewhere in the next block, jumping and exulting in their leaves, maybe catching fire).

It was a mistake to change this, smoke or no smoke, carbon dioxide and the greenhouse effect or whatever; it was a loss to give up the burning of autumn leaves. Now, in our haste to protect the environment (which is us, when you get down to it), we rake them up and cram them into great black plastic bags, set out at the curb like wrapped corpses, carted away by the garbage truck to be buried somewhere or dumped in the sea or made into fuel or alcohol or whatever it is they do with autumn leaves these days. We should be giving them back to the children to burn.

My Magical Metronome

I woke up, late one Friday night, feeling like the Long Island Railroad thumping at top speed over a patch of bad roadbed. Doctor-fashion, I took my pulse and found it too fast to count accurately. I heaved out of bed and sat in a chair, gloomy, wondering what next. A while later the train slowed down, nearly stopped, and my pulse rate had suddenly dropped to 35. I decided to do some telephoning.

Next thing I knew, I was abed in the intensive care unit of the hospital down the street, intravenous tubes in place, wires leading from several places on my chest and from electrodes on my arms and legs, lights flashing from the monitor behind my bed. If I turned my head sharply I could see the bouncing lines of my electrocardiogram, a totally incomprehensible graffito, dropped beats, long stretches of nothing followed by what looked like exclamation points.

The handwriting on the wall, I thought. And illiterate at that.

Now it was Sunday, late afternoon, the monitor still jumpy, alarm lights still signaling trouble, all the usual drugs for restoring cardiac rhythm having been tried, and handwriting still a scrawl. The cardiovascular surgeon at the foot of my bed was explaining that it would have to be a pacemaker, immediately, Sunday late afternoon. What did I think?

What I thought, and then said, was that this was one of the things about which a man is not entitled to his own opinion. Over to you, I said.

About an hour later I was back from the operating theater. Theater is right; the masked surgeon center stage, wonderfully lit, several colleagues as appreciative audience, me as the main prop. The denouement was that famous *deus ex machina* being inserted into the prop's chest wall, my gadget now, my metronome. Best of all, my heart rate an absolutely regular, dependable, reliable 70, capable of speeding up on demand but inflexibly tuned to keep it from dropping below 70. The battery guaranteed to last seven years or thereabouts before needing changing. Plenty of time to worry about that, later on.

Home in a couple of days, up and around doing whatever I felt like, up and down stairs, even pushing furniture from one place to another, then back to work.

Afterthought:

A new, unwarranted but irrepressible kind of vanity. I had come into the presence of a technological marvel, namely me. To be sure, the pacemaker is a wonderful miniature piece of high technology, my friend the surgeon a skilled worker in high technology, but the greatest of wonders is my own pump, my myocardium, capable of accepting electronic instructions from that small black box and doing exactly what it is told. I am exceedingly pleased with my machine-tooled, obedient, responsive self. I would never have thought I had it in me, but now that I have it in me, ticking along soundlessly, flawlessly, I am subject to waves of pure vanity.

Another surprise:

I do not want to know very much about my new technology. I do not even want to have the reasons for needing it fully explained to me. As long as it works, and it does indeed, I prefer to be as mystified by it as I can. This is a surprise. I would have thought that as a reasonably intelligent doctor-patient I would be filled with intelligent, penetrating questions, insisting on comprehending each step in the procedure, making my own decisions, even calling the shots. Not a bit of it. I turn out to be the kind of patient who doesn't want to have things explained, only to have things looked after by the real professionals. Just before I left the hospital, the cardiologist brought me a manila envelope filled with reprints, brochures, the pacemaker manufacturer's instructions for physicians listing all the indications, warnings, the things that might go wrong. I have the envelope somewhere, on a closet shelf I think, unexamined. I haven't, to be honest, the faintest idea how a pacemaker works, and I have even less curiosity.

This goes against the wisdom of the times, I know. These days one reads everywhere, especially in the popular magazines, that a patient should take more responsibility, be more assertive, insist on second and third opinions, and above all have everything fully explained by the doctor or, preferably, the doctors, before submitting to treatment. As a physician, I used to think this way myself, but now, as a successful patient, I feel different. Don't explain it to me, I say, go ahead and fix it.

I suppose I should be feeling guilty about this. In a way I do, for I have written and lectured in the past about medicine's excessive dependence on technology in general, and the resultant escalation in the cost of health care. I have been critical of what I called "halfway technologies," designed to shore things up and keep flawed organs functioning beyond their appointed time. And here I am, enjoying precisely this sort of technology, eating my words.

Pacemakers have had a bad press recently, with stories about overutilization, kited prices, kickbacks to doctors and hospitals, a scandal. Probably the stories, some of them anyway, are true. But I rise to the defense of the gadget itself, in which I now have so personal a stake. If anyone had tried to tell me, long ago when I was a medical student, that the day would come when a device the size of a cigarette lighter could be implanted permanently over the heart, with wires extending to the interior of the ventricle, dominating the heart's conduction system and regulating the rhythm with perfection, I would have laughed in his face. If then he had told me that this would happen one day to me, I would have gotten sore. But here it is, incomprehensible, and I rather like it.

On Speaking of Speaking

There is nothing at all wrong with the English language, so far as I can see, but that may only be because I cannot see ahead. If I were placed in charge of it, as chairman, say, of a National Academy for the Improvement of Language, I would not lay a finger on English. It suits every need that I can think of: flexibility, clarity, subtlety of metaphor, ambiguity wherever ambiguity is needed (which is more often than is generally acknowledged), and most of all changeability. I like the notion of a changing language. As a meliorist, I am convinced that all past changes were for the better; I have no doubt that today's English is a considerable improvement over Elizabethan or Chaucerian talk, and miles ahead of Old English. By now the language has reached its stage of ultimate perfection, and I'll be satisfied to have it this way forever.

But I know I'm wrong about this. English is shifting and

changing before our eyes and ears, beyond the control of all individuals, committees, academies, and governments. The speakers of earlier versions undoubtedly felt the same satisfaction with their speech in their time. Chaucer's generation, and all the generations before, could not have been aware of any need to change or improve. Montaigne was entirely content with sixteenth-century French and obviously delighted by what he could do with it. Long, long ago, the furthermost ancestors of English speech must have got along nicely in Proto-Indo-European without a notion that their language would one day vanish.

"Vanish" is the wrong word anyway for what happened. The roots of several thousand Indo-European words are still alive and active, tucked up neatly like symbionts inside other words in Greek, Latin, and all the Germanic tongues, including English. Much of what we say to each other today, in English, could be interpreted as Greek with an Indo-European accent. Three or four centuries from now, it is probable that today's English will be largely incomprehensible to everyone except the linguistic scholars and historians.

The ancient meanings of the Indo-European roots are sometimes twisted around, even distorted beyond recognition, but they are still there, resonating inside, reminding. The old root *gheue*, meaning simply to call, became *gudam* in Germanic and then "God" in English. *Meug* was a root signifying something damp and slippery, and thousands of years later it turned into "meek" in proper English and "mooch" in slang, also "schmuck." *Bha* was the Indo-European word for speaking, becoming *phanai* in Greek with the same meaning, then used much later for our most fundamental word indicating the inability to speak: "infancy." *Ster* was a root meaning to stiffen; it became *sterban* in Germanic and *steorfan* in Old English, meaning to die, and then turned into "starve" in our speech.

The changes in language will continue forever, but no one knows for sure who does the changing. One possibility is that children are responsible. Derek Bickerton, professor of linguistics at the University of Hawaii, explores this in his book *Roots of Language*. Sometime around 1880, a language catastrophe occurred in Hawaii when thousands of immigrant workers were brought to the islands to work for the new sugar industry. These people, speaking Chinese, Japanese, Korean, Portuguese, and various Spanish dialects, were unable to communicate with one another or with the native Hawaiians or the dominant English-speaking owners of the plantations, and they first did what such mixed-language populations have always done: they spoke Pidgin English (a corruption of "business English"). A pidgin is not really a language at all, more like a set of verbal signals used to name objects but lacking the grammatical rules needed for expressing thought and ideas. And then, within a single generation, the whole mass of mixed peoples began speaking a totally new tongue: Hawaiian Creole. The new speech contained ready-made words borrowed from all the original tongues, but bore little or no resemblance to the predecessors in the rules used for stringing the words together. Although generally regarded as a "primitive" language, Hawaiian Creole was constructed with a highly sophisticated grammar. Professor Bickerton's great discovery is that this brand-new speech could have been made only by the children. There wasn't time enough to allow for any other explanation. Soon after the influx of workers in 1880 the speech was Hawaiian Pidgin, and within the next twenty-five or thirty years the accepted language was Creole. The first immigrants, the parents who spoke Pidgin, could not have made the new language and then taught it to the children. They could not themselves understand Creole when it appeared. Nor could the adult English speakers in charge of the place either speak or comprehend Creole. According to Bickerton's research, it simply had to have been the work of children, crowded together, jabbering away at each other, playing.

Bickerton cites this historic phenomenon as evidence, incontrovertible in his view, for the theory that language is a biological, innate, genetically determined property of human beings, driven by a center or centers in the brain that code out grammar and syntax. His term for the gift of speech is "bioprogram." The idea confirms and extends the proposal put forward by Noam Chomsky, almost three decades ago, that human beings are unique in their possession of brains equipped for generating grammar. But the most fascinating aspect of the new work is its evidence that children—and probably very young children at that—are able to construct a whole language, working at it together, or more likely *playing* at it together.

It should make you take a different view of children, eliciting something like awe. We have always known that childhood is the period in which new languages as well as one's own can be picked up quickly and easily. The facility disappears in most people around the time of adolescence, and from then on the acquisition of a new language is hard, slogging labor. Children are gifted at it, of course. But it requires a different order of respect to take in the possibility that children make up languages, change languages, perhaps have been carrying the responsibility for evolving language from the first human communication to twentieth-century speech. If it were not for the children and their special gift we might all be speaking Indo-European or Hittite, but here we all are, speaking several thousand different languages and dialects, most of which would be incomprehensible to the human beings on earth just a few centuries back.

Perhaps we should be paying serious attention to the possible role played by children in the origin of speech itself. It is of course not known when language first appeared in our species, and it is pure guesswork as to how it happened. One popular guess is that at a certain stage in the evolution of the human skull, and of the brain therein, speech became a possibility in a few mutant individuals. Thereafter, these intellectual people and their genes outcompeted all their speechless cousins, and natural selection resulted in *Homo sapiens*. This notion would require the recurrence of the same mutation in many different, isolated communities all around the globe, or else one would have to assume that a lucky few speakers managed to travel with remarkable agility everywhere on earth, leaving their novel genes behind.

Another possibility, raised by the new view of children and speech, is that human language did not pop up as a special mutation, but came into existence as a latent property of all human brains at some point in the evolution of the whole species. The environment required for expression of the brain centers involved in the process was simply children, enough children crowded together in circumstances where they could spend a lot of time playing together. A critical mass of children in a sufficiently stable society could have been achieved whenever large enough numbers of families settled down to live in close quarters, as may have happened long ago in the tribal life of hunters and gatherers or in the earliest agricultural communities.

It makes an interesting scenario. The adults and wise elders of the tribe, sitting around a fire speaking a small-talk pidgin, pointing at one thing or another and muttering isolated words. No syntax, no strings of words, no real ideas, no metaphors. Somewhere nearby, that critical mass of noisy young children, gabbling and shouting at each other, their voices rising in the exultation of discovery, talking, talking, and forever thereafter never stopping.

Seven Wonders

A while ago I received a letter from a magazine editor inviting me to join six other people at dinner to make a list of the Seven Wonders of the Modern World, to replace the seven old, out-of-date Wonders. I replied that I couldn't manage it, not on short order anyway, but still the question keeps hanging around in the lobby of my mind. I had to look up the old biodegradable Wonders, the Hanging Gardens of Babylon and all the rest, and then I had to look up that word "wonder" to make sure I understood what it meant. It occurred to me that if the magazine could get any seven people to agree on a list of any such seven things you'd have the modern Seven Wonders right there at the dinner table.

Wonder is a word to wonder about. It contains a mixture of messages: something marvelous and miraculous, surprising, raising unanswerable questions about itself, making the observer wonder, even raising skeptical questions like, "I *wonder* about that." Miraculous and marvelous are clues; both words come from an ancient Indo-European root meaning simply to smile or to laugh. Anything wonderful is something to smile in the presence of, in admiration (which, by the way, comes from the same root, along with, of all telling words, "mirror").

I decided to try making a list, not for the magazine's dinner party but for this occasion: seven things I wonder about the most.

I shall hold the first for the last, and move along.

My Number Two Wonder is a bacterial species never seen on the face of the earth until 1982, creatures never dreamed of before, living violation of what we used to regard as the laws of nature, things literally straight out of Hell. Or anyway what we used to think of as Hell, the hot unlivable interior of the earth. Such regions have recently come into scientific view from the research submarines designed to descend twenty-five hundred meters or more to the edge of deep holes in the sea bottom, where open vents spew superheated seawater in plumes from chimneys in the earth's crust, known to oceanographic scientists as "black smokers." This is not just hot water, or steam, or even steam under pressure as exists in a laboratory autoclave (which we have relied upon for decades as the surest way to destroy all microbial life). This is extremely hot water under extremely high pressure, with temperatures in excess of 300 degrees centigrade. At such heat, the existence of life as we know it would be simply inconceivable. Proteins and DNA would fall apart, enzymes would melt away, anything alive would die instantaneously. We have long since ruled out the possibility of life on Venus because of that planet's comparable temperature; we have ruled out the possibility of life in the earliest years of this planet, four billion or so years ago, on the same ground.

B. J. A. Baross and J. W. Deming have recently discovered the presence of thriving colonies of bacteria in water fished directly from these deep-sea vents. Moreover, when brought to the surface, encased in titanium syringes and sealed in pressurized chambers heated to 250 degrees centigrade, the bacteria not only survive but reproduce themselves enthusiastically. They can be killed only by chilling them down in boiling water.

And yet they look just like ordinary bacteria. Under the electron microscope they have the same essential structure— cell walls, ribosomes, and all. If they were, as is now being suggested, the original archebacteria, ancestors of us all, how did they or their progeny ever learn to cool down? I cannot think of a more wonderful trick.

My Number Three Wonder is *oncideres*, a species of beetle encountered by a pathologist friend of mine who lives in Houston and has a lot of mimosa trees in his backyard. This beetle is not new, but it qualifies as a Modern Wonder because of the exceedingly modern questions raised for evolutionary biologists about the three consecutive things on the mind of the female of the species. Her first thought is for a mimosa tree, which she finds and climbs, ignoring all other kinds of trees in the vicinity. Her second thought is for the laying of eggs, which she does by crawling out on a limb, cutting a longitudinal slit with her mandible and depositing her eggs beneath the slit. Her third and last thought concerns the welfare of her offspring; beetle larvae cannot survive in live wood, so she backs up a foot or so and cuts a neat circular girdle all around the limb, through the bark and down into the cambium. It takes her eight hours to finish this cabinetwork. Then she leaves and where she goes I do not know. The limb dies from the girdling, falls to the ground in the next breeze, the larvae feed and grow into the next generation, and the questions lie there unanswered. How on earth did these three linked thoughts in her mind evolve together in evolution? How could any one of the three become fixed as beetle behavior by itself, without the other two? What are the odds favoring three totally separate bits of behavior—liking a particular tree, cutting a slit for eggs, and then girdling the limb—happening together by random chance among a beetle's genes? Does this smart beetle know what she is doing? And how did the mimosa tree enter the picture in its evolution? Left to themselves, unpruned, mimosa trees have a life expectancy of twenty-five to thirty years. Pruned each year, which is what the beetle's girdling labor accomplishes, the tree can flourish for a century. The mimosa-beetle relationship is an elegant example of symbiotic partnership, a phenomenon now recognized as pervasive in nature. It is good for us to have around on our intellectual mantelpiece such creatures as this insect and its friend the tree, for they keep reminding us how little we know about nature.

The Fourth Wonder on my list is an infectious agent known as the scrapie virus, which causes a fatal disease of the brain in sheep, goats, and several laboratory animals. A close cousin of scrapie is the C-J virus, the cause of some cases of senile dementia in human beings. These are called "slow viruses," for the excellent reason that an animal exposed to infection today will not become ill until a year and a half or two years from today. The agent, whatever it is, can propagate itself in abundance from a few infectious units today to more than a billion next year. I use the phrase "whatever it is" advisedly. Nobody has yet been able to find any DNA or RNA in the scrapie or C-J viruses. It may be there, but if so it exists in amounts too small to detect. Meanwhile, there is plenty of protein, leading to a serious proposal that the virus may indeed be *all* protein. But protein, so far as we know, does not replicate itself all by itself, not on this planet anyway. Looked at this way, the scrapie agent seems the strangest thing in all biology and, until someone in some laboratory figures out what it is, a candidate for Modern Wonder.

My Fifth Wonder is the olfactory receptor cell, located in the epithelial tissue high in the nose, sniffing the air for clues to the environment, the fragrance of friends, the smell of leaf smoke, breakfast, nighttime and bedtime, and a rose, even, it is said, the odor of sanctity. The cell that does all these things, firing off urgent messages into the deepest parts of the brain, switching on one strange unaccountable memory after another, is itself a proper brain cell, a certified neuron belonging to the brain but miles away out in the open air, nosing around the world. How it manages to make sense of what it senses, discriminating between jasmine and anything else non-jasmine with infallibility, is one of the deep secrets of neurobiology. This would be wonder enough, but there is more. This population of brain cells, unlike any other neurons of the vertebrate central nervous system, turns itself over every few weeks; cells wear out, die, and are replaced by brand-new cells rewired to the same deep centers miles back in the brain, sensing and remembering the same wonderful smells. If and when we reach an understanding of these cells and their functions, including the moods and whims under their governance, we will know a lot more about the mind than we do now, a world away.

Sixth on my list is, I hesitate to say, another insect, the termite. This time, though, it is not the single insect that is the Wonder, it is the collectivity. There is nothing at all wonderful about a single, solitary termite, indeed there is really no such creature, functionally speaking, as a lone termite, any more than we can imagine a genuinely solitary human being; no such thing. Two or three termites gathered together on a dish are not much better; they may move about and touch each other nervously, but nothing happens. But keep adding more termites until they reach a critical mass, and then the miracle begins. As though they had suddenly received a piece of extraordinary news, they organize in platoons and begin stacking up pellets to precisely the right height, then turning the arches to connect the columns, constructing the cathedral and its chambers in which the colony will live out its life for the decades ahead, air-conditioned and humidity-controlled, following the chemical blueprint coded in their genes, flawlessly, stone-blind. They are not the dense mass of individual insects they appear to be; they are an organism, a thoughtful, meditative brain on a million legs. All we really know about this new thing is that it does its architecture and engineering by a complex system of chemical signals.

The Seventh Wonder of the modern world is a human child, any child. I used to wonder about childhood and the evolution of our species. It seemed to me unparsimonious to keep expending all that energy on such a long period of vulnerability and defenselessness, with nothing to show for it, in biological terms, beyond the feckless, irresponsible pleasure of childhood. After all, I used to think, it is one sixth of a whole human life span! Why didn't our evolution take care of that, allowing us to jump catlike from our juvenile to our adult (and, as I thought) productive stage of life? I had forgotten about language, the single human trait that marks us out as specifically human, the property that enables our survival as the most compulsively, biologically, obsessively social of all creatures on earth, more interdependent and interconnected even than the famous social insects. I had forgotten that, and forgotten that children *do* that in childhood. Language is what childhood is for.

There is another related but different creature, nothing like so wonderful as a human child, nothing like so hopeful, something to worry about all day and all night. It is *us*, aggregated together in our collective, critical masses. So far, we have learned how to be useful to each other only when we collect in small groups—families, circles of friends, once in a while (although still rarely) committees. The drive to be useful is encoded in our genes. But when we gather in very large numbers, as in the modern nation-state, we seem capable of levels of folly and self-destruction to be found nowhere else in all of Nature.

As a species, taking all in all, we are still too young, too juvenile, to be trusted. We have spread across the face of the earth in just a few thousand years, no time at all as evolution clocks time, covering all livable parts of the planet, endangering other forms of life, and now threatening ourselves. As a species, we have everything in the world to learn about living, but we may be running out of time. Provisionally, but only provisionally, we are a Wonder.

And now the first on my list, the one I put off at the beginning of making a list, the first of all Wonders of the modern world. To name this one, you have to redefine the world as it has indeed been redefined in this most scientific of all centuries. We named the place we live in the *world* long ago, from the Indo-European root *wiros*, which meant man. We now live in the whole universe, that stupefying piece of expanding geometry. Our suburbs are the local solar system, into which, sooner or later, we will spread life, and then, likely, beyond into the galaxy. Of all celestial bodies within reach or view, as far as we can see, out to the edge, the most wonderful and marvelous and mysterious is turning out to be our own planet earth. There is nothing to match it anywhere, not yet anyway.

It is a living system, an immense organism, still developing, regulating itself, making its own oxygen, maintaining its own temperature, keeping all its infinite living parts connected and interdependent, including us. It is the strangest of all places, and there is everything in the world to learn about it. It can keep us awake and jubilant with questions for millennia ahead, if we can learn not to meddle and not to destroy. Our great hope is in being such a young species, thinking in language only a short while, still learning, still growing up.

We are not like the social insects. They have only the one way of doing things and they will do it forever, coded for that way. We are coded differently, not just for binary choices, *go* or *no-go*. We can go four ways at once, depending on how the air feels: *go*, *no-go*, but also *maybe*, plus *what the hell let's give it a try*. We are in for one surprise after another if we keep at it and keep alive. We can build structures for human society never seen before, thoughts never thought before, music never heard before.

Provided we do not kill ourselves off, and provided we can connect ourselves by the affection and respect for which I believe our genes are also coded, there is no end to what we might do on or off this planet.

At this early stage in our evolution, now through our infancy and into our childhood and then, with luck, our growing up, what our species needs most of all, right now, is simply a future.

The Artificial Heart

A short while ago, I wrote an essay in unqualified praise of that technological marvel the pacemaker, celebrating the capacity of this small ingenious device to keep a flawed human heart working beyond what would otherwise have been its allotted time. I had no reservations about the matter: here was an item of engineering that ranks as genuine high technology, a stunning example of what may lie ahead for applied science in medicine.

And then, out of Salt Lake City, came the news of the artificial heart, a functioning replacement for the whole organ, far outclassing anything like my miniature metronome, a science-fiction fantasy come true.

My reaction to the first headlines, on the front pages of all the papers and in the cover stories of the newsmagazines, was all admiration and pleasure. A triumph, to be sure. No question about it.

But then the second thought, and the third and fourth thoughts, dragging their way in and out of my mind leaving one worry after another: What happens now? If the engineers keep at it, as they surely will, and this remarkable apparatus is steadily improved—as I'm sure it can be—so that we end up without the need for that cart with its compressors and all those hoses, an entirely feasible replacement for anyone's failing heart, what then? The heart disease called cardiomyopathy, for which the initial device was employed, is a relatively uncommon, obscure ailment, entailing very high cost, but for a limited number of patients; no great strain on the national economy. But who says that an artificial heart will be implanted only in patients with a single, rare form of intractable heart failure? What about the hundreds of thousands of people whose cardiac muscles have been destroyed by coronary atherosclerosis and who must otherwise die of congestive heart failure? Who will decide that only certain patients, within certain age groups, will be selected for this kind of lifesaving (or at least life-prolonging) technology? Will there be committees, sitting somewhere in Washington, laying out national policy? How can Congress stay out of the problem, having already set up a system for funding the artificial kidney (with runaway costs already far beyond the original expectations and no end in sight)? And where is the money to come from, at a time when every penny of taxpayers' money for the health-care system is being pinched out of shape?

I conclude that the greatest potential value of the successful artificial heart is, or ought to be, its power to convince the government as well as the citizenry at large that the nation simply must invest more money in basic biomedical research.

We do not really understand the underlying mechanism of cardiomyopathies at all, and we are not much better off at comprehending the biochemical events that disable the heart muscle or its valves in other more common illnesses. But there are clues enough to raise the spirits of people in a good many basic science disciplines, and any number of engrossing questions are at hand awaiting answers. The trouble is that most of the good questions that may lead, ultimately, to methods for prevention (for example, the metabolism and intimate pathologic changes in a failing myocardium, the possible roles of nutrition, viral infection, blood-clotting abnormalities, hypertension, life-style, and other unknown factors) are all long-range questions, requiring unguessable periods of time before the research can be completed. Nor can the outcome of research on any particular line be predicted in advance; whatever turns up as the result of science is bound to be new information. There can be no guarantee that the work will turn out to be useful. It can, however, be guaranteed that if such work is not done we will be stuck forever with this insupportably expensive, ethically puzzling, halfway technology, and it is doubtful that we can long afford it.

We are in a similar fix for the other major diseases, especially the chronic ones affecting the aging population. Although nothing so spectacular as the artificial heart has emerged for the treatment of stroke, or multiple sclerosis, or dementia, or arthritis, or diabetes, or cirrhosis, or advanced cancer, or the others on the list, the costs of whatever therapy we do possess continue to escalate at a terrifying rate.

Soon we will be spending more than 10 percent of the GNP on efforts to cope with such chronic health problems. The diseases are all comparable in at least one respect: it cannot be promised that scientific research will solve them, but it can be firmly predicted that without research there is no hope at all of preventing or getting rid of them.

The artificial heart could, with better science and a lot of luck, turn out to be, one day or other, an interesting kind of antique, similar in its historical significance to the artificial lung and the other motor-driven prosthetic devices that were in the planning stage just before the development of the Salk vaccine and the virtual elimination of poliomyelitis. Or the complex and costly installations for lung surgery that were being planned for the state sanatoriums just before the institutions themselves were closed by the development of effective chemotherapy for tuberculosis.

The biological revolution of the past three decades has placed at the disposal of biomedical science an array of research techniques possessing a power previously unimaginable. It should be possible, henceforth, to ask questions about the normal and pathologic functions of cells and tissues at a very profound level, questions that could not even have been thought up as short a time ago as ten years. We should be thinking more about this new turn of events while meditating on the meaning of the artificial heart.

Things Unflattened by Science

In one of her Norton Lectures at Harvard in 1980, Helen Gardner had some sharply critical things to say about criticism, particularly about the reductionist tendencies of contemporary literary criticism, and especially about the new New Criticism out of France known as deconstructionism, the reductionist fission of poetry, not line by line but word by word, particle by particle. She was worried about the new dogma that the poem itself cannot possess any meaning whatever, beyond the random insights brought to the words by the reader, the observer. The only reality to be perceived in a line of verse is a stochastic reality arranged by the observer, not by the creator of the line. Miss Gardner is dismayed by this affront to literature. "It marks," she writes, "a real loss of belief in the value of literature and of literary study, . . . dignified and partly justified by being linked with a universal skepticism about the possibility of any real knowledge of the universe we live in or any true understanding of the world of our daily experience." The "indeterminacy of literary texts," she says, "is part of the indeterminacy of the world."

Joan Peyser, in an introduction to the new edition of her ten-year-old book on modern music, expresses a similar level of dismay at what is happening to contemporary music. She writes, "The lessening of greatness in the music of modern times can be traced to Darwin, Marx, Einstein and Freud"; she adds, "the dissemination of their theories propelled everything hidden into the light; analysis annihilates mystery."

The geneticist C. H. Waddington asserted in his book on modern art that some of the earliest manifestations of abstract expression in modern painting, notably the work of Kandinsky and his followers, came from a feeling of hostility toward early twentieth-century physics. Kandinsky believed that scientists were "capable only of recognizing those things that can be weighed and measured."

Annie Dillard, writing about the impact of modern physics on modern fiction, in a wonderful book on criticism entitled *Living by Fiction*, says, "nothing is more typical of modernist fiction than its shattering of narrative line. . . . The use of narrative collage is particularly adapted to twentieth-century treatments of time and space . . . a flattened landscape. . . . Events do not trigger other events at all; instead, any event is possible. . . . The world is an undirected energy; it is an infinite series of random possibilities." "This," she continues, "is the fiction of quantum mechanics," and she doesn't care much for it. She believes that there is meaning in the world, but concludes that the lyric poets are the best equipped of all of us to find it.

I wish the humanists, wherever they are—the artists, writers, poets, critics, and musicians (most of all the musicians)—would leave physics alone for a while and begin paying more attention to biology. Personally, having read my way through a long shelf of books written by physicists for nonmathematicians like me, I have given up looking for the meaning, any meaning at all, in the worlds of very small or very large events. I've become convinced that any effort to insert mysticism into quantum mechanics, or to get mysticism out of it, or indeed to try to force new meanings into the affairs of the everyday, middle-sized world, is not for me. There are some things about which it is not true to say that every man has a right to his own opinion. I do not have the right to an opinion about acausality in the small world, or about black holes or other universes beyond black holes in the large world, for I cannot do the mathematics. Physics, deep and beautiful physics, can be spoken only in pure, unaccented mathematics, and no other language exists for expressing its meaning, not yet anyway. Lacking the language, I concede that it is none of my business, and I am giving up on it.

Biology is something else again, another matter, quite another matter indeed, in fact very likely another form, or at least another aspect of matter, probably not glimpsed, or anyway not yet glimpsable, by the mathematics of quantum physics.

One big difference is that biology, being a more difficult science, has lagged behind, so far behind that we have not yet reached the stage of genuine theory—in the predictive sense in which theoretical physics drives that field along. Biologists are still principally engaged in making observations and collecting facts, trying wherever possible to relate one set of facts to another but still lacking much of a basis for grand unifying theories. Evolution is about as close as we have come, and it is certainly a grand and sweeping concept, but more like a wonderful puzzle, filled with bits of information waiting for more bits before the whole matter can be fitted together. It remains, necessarily, an intensely reductionist field in science, requiring the scrutiny of endless details, and then the details of the details, before it will become possible to see a large, clear picture of the whole orderly process, and it will need decades of work, perhaps centuries, before we can stand back for a long look. It may even be that some of the information lies forever beyond our grasp because of the sheer age and volume of planetary life and the disappearance from the record of so many crucial forms, crucial for comprehending the course of events.

In fact, we can look back only a relatively short distance. Up until the 1950s, the fossil record, on which the most solid parts of the structure of evolutionary theory were based, provided a fairly close look at only the last five hundred million years or so. We now know, from the work of Barghoorn, Cloud, Schopf, and others, that there is a period of at least three billion years of life about which we know very little, and for most of that time the sole occupants of the earth were the prokaryotes—bacteria and, I have no doubt, their resident viruses. We tend to use words such as "early" and "primitive" for such creatures, as though we members of the eukaryote world, possessing nucleated cells and on the way to making brains for ourselves, comprise a qualitatively different and vastly superior form of life. We tend sometimes even to dismiss four-fifths of the earth's life span as a long, dull prologue to the *real* events in evolution, nothing but featureless, aimless bacteria around, waiting for the real show to begin.

It was probably not like that at all. Leave aside the excitement when the very first successful cell appeared, membranes, nucleic acid, ribosomes, proteins, and all, somewhere in a quiet pool, maybe in the aftermath of a lightning storm, maybe from a combination of energy sources: the sun, ionizing radiation and volcanic heat. It can be told as a plausible story, easy to imagine, for all the necessary chemical building blocks were at hand (or came to hand) during the first billion years, and it should no longer come as a surprise that beautifully formed bacterial fossils exist in rocks 3.5 billion years old. I wish, by the way, that we had set up a better term—a *nicer* term—than "primordial soup" for the nutrients and clay surfaces in those early waters of the earth. "Soup" is somehow too dismissive a word for a state of affairs so immensely important, more like the role of the yolk in a fertilized egg (although that doesn't sound much better). Maybe it is an unexplored tradition in the language of science to flatten out the prose for really huge events: what may be turning out to be the most profound and subtle of all mechanisms in evolutionary genetics is now known, flatly and familiarly, as "jumping genes."

The first cell to appear on the planet was in all probability just that: a single first cell, capable of replicating itself, and a creature of great theoretical interest. But the events that followed over the next 2.5 billion or so years seem to me even more fascinating. It is entirely possible that the stretch of time was needed for the progeny of the first cell to learn virtually everything essential for getting on in a closed ecosystem. Long before the first great jump could be taken—the transformation of prokaryotes to eukaryotes around a billion years ago—a great many skills had to be acquired.

During those years, the life of the earth was of course made up of vast numbers of individual cells, each one replicating on its own, but it would have seemed to an outside observer more like a tissue, the differentiated parts of a huge organism, than a set of discrete beings. In most places, and in the algal mats that covered much of the earth's surface for a very long time, the microorganisms arranged themselves in neatly aligned layers, feeding one another in highly specialized ways and developing the mechanisms for cooperation and coordination that, I believe, have characterized the biosphere ever since.

Chemical messengers of precision and subtlety evolved during this stage, used no doubt for the allocation of space and the encouragement (or discouragement) of replication by neighboring microorganisms. Some of these chemical signals are still with us, but now they are emitted from specialized cells in the tissues of higher organisms, functioning as hormones. Insulin, for example, or a protein very similar to insulin with similar properties, is produced by strains of that famous and ancient bacterium, *E. coli.* Other bacteria are known to make a substance similar to human chorionic gonadotropin. Later, when protozoa and fungi evolved from their ancestral prokaryotes, they came equipped with

ACTH, insulin, and growth hormone, all similar to their modern counterparts.

Moreover, the life of the planet began the long, slow process of modulating and regulating the physical conditions of the planet. The oxygen in today's atmosphere is almost entirely the result of photosynthetic living, which had its start with the appearance of blue-green algae among the microorganisms. It was very likely this first step—or evolutionary jump—that led to the subsequent differentiation into eukaryotic, nucleated cells, and there is almost no doubt that these new cells were pieced together by the symbiotic joining up of prokaryotes. The chloroplasts in today's green plants, which capitalize on the sun's energy to produce the oxygen in our atmosphere, are the lineal descendants of ancient blue-green algae. The mitochondria in all our cells, which utilize the oxygen for securing energy from plant food, are the progeny of ancient oxidative bacteria. Collectively, we are still, in a fundamental sense, a tissue of microbial organisms living off the sun, decorated and ornamented these days by the elaborate architectural structures that the microbes have constructed for their living quarters, including seagrass, foxes, and of course ourselves.

We can imagine three worlds of biology, corresponding roughly to the three worlds of physics: the very small world now being explored by the molecular geneticists and virologists, not yet as strange a place as quantum mechanics but well on its way to strangeness; an everyday, middle-sized world where things are as they are; and a world of the very large, which is the whole affair, the lovely conjoined biosphere, the vast embryo, the closed ecosytem in which we live as working parts, the place for which Lovelock and

Margulis invented the term "Gaia" because of its extraordinary capacity to regulate itself. This world seems to me an even stranger one than the world of very small things in biology: it looks like the biggest organism I've ever heard of, and at the same time the most delicate and fragile, exactly the delicate and fragile creature it appeared to be in those first photographs taken from the surface of the moon. It is at this level of things that I find meaning in Wallace Stevens, although I haven't any idea that Stevens intended this in his "Man with the Blue Guitar": "they said, 'You have a blue guitar,/ you do not play things as they are.'/ The man replied, 'Things as they are/ are changed upon the blue guitar.' " It is a long poem, alive with ambiguities, but it can be read, I think, as a tale of the earth itself.

Some biologists dislike the Lovelock-Margulis view of things, although they agree that the regulatory homeostasis of earth's life exists as a real phenomenon. They dislike the term "Gaia," for one thing, because of its possible religious undertones, and they dislike the notion of design that seems implicit—although one way out of that dilemma is to call the arrangement a "System" and then assert that this is the only way that complex "Systems" can survive, by endless chains of regulatory messages and intricate feedback loops. It is not necessary, in accounting for the evolution and now the stability of the earth's atmosphere, to suggest that evolution itself can plan ahead; all you need assume is the existence of close linkages of interdependency involving all existing forms of life, after the fashion of an organism. Finally, it is not a view of things, as has been claimed, that is likely to relieve human beings of any feeling of responsibility for the environment, backing them off from any concern for the whole place, on grounds that it runs itself and has done so, implacably, since long before we arrived on the scene. To the contrary, I should think it would have just the opposite effect, imposing a new feeling of anxiety for the environment everywhere. If you become convinced that you exist as a part of something that is itself alive, you are more likely to take pains not to do damage to the other vital parts around you.

Anyway, it seems to me a notion in biology not to be dismissed lightly, and requiring a great deal more thought and a lot more science. Part of the science can be done best by the technologies developed for space exploration. One thing discovered since NASA began its work, on which I should think almost everyone would now agree, is that by far the most interesting, engrossing, and puzzling object in the solar system—maybe in the whole galaxy—is our own planet. It needs more research, huge-scale and at the same time delicate, highly reductionist work, but in the meantime it is there for the humanists to think about, something new and amiable, a free gift from science and high technology, a nice piece of bewilderment for the poets, an instruction in humility for all the rest of us.

In the everyday middle-sized world where I live, biology has only begun to work. Medicine, the newest and youngest of all the sciences, bobs along in the wake of biology, indeed not yet sure that it *is* all that much a science, but certain that if there is to be a scientific future for medicine it can come only from basic biomedical research. I'm not sure who invented that convenient hybrid word "biomedical." I think it was someone from my professional side, wanting to lay claims on respectable science by the prefix "bio"; but it could as well have been a pure biologist wanting the suffix "medical" as a way to lay hands on NIH grants. Whichever, it is a nice word and it contains the truth: medicine is a branch of biological science for the long-term future.

This means that I am entitled, as a physician, to ask my biologist friends to answer a range of questions that are not yet perceived as an immediate part of my own bailiwick, just as they can expect me and my colleagues to turn up some quick answers to problems like cancer, coronary disease, schizophrenia, heartburn, whatever. Indeed, the only question I am inclined to turn aside as being impossible to respond to happens to be the one most often raised these days, not just by my biologist friends but by everyone: the question about stress, how to avoid stress, prevent stress, allay stress. I refuse to have anything to do with this matter, having made up my mind, from everything I have read or heard about it in recent years, that what people mean by stress is simply the condition of being human, and I will not recommend any meddling with that, by medicine or any other profession.

But I digress. What I wish to get at is an imaginary situation in which I am allowed three or four questions to ask the world of biomedical science to settle for me by research, as soon as possible. Can I make a short list of top-priority puzzles, things I am more puzzled by than anything else? I can.

First, I want to know what goes on in the mind of a honeybee. Is it true, as is often asserted, that a bee is simply a small, neatly assembled robot, capable only of behaving in ways for which the bee is programmed by instructions in bee DNA, or is something else going on? In short, does a bee know what is going on in its mind when it navigates its way to distant food sources and back to the hive, using polarized sunlight and the tiny magnet it carries as a navigational aid? Or is the bee just a machine, unable to do its mathematics and dance its language in any other way? To use Donald Griffin's term, does a bee have "awareness"; or to use a phrase I like better, can a bee think and imagine?

There is an experiment for this, or at least an observation made long ago by Karl von Frisch and more recently confirmed by James Gould at Princeton. Biologists who wish to study such things as bee navigation, language, and behavior in general have to train their bees to fly from the hive to one or another special place. To do this, they begin by placing a source of sugar very close to the hive so that the bees (considered by their trainers to be very dumb beasts) can learn what the game is about. Then, at regular intervals, the dish or whatever is moved progressively farther and farther from the hive, in increments of about 25 percent at each move. Eventually, the target is being moved a hundred feet or more at a jump, very far from the hive. Sooner or later, while this process is going on, the biologist shifting the dish of sugar will find that his bees are out there waiting for him, precisely where the next position had been planned. This is an uncomfortable observation to make, harder still to explain in conventional terms: Why would bees be programmed for such behavior in their evolution? Flowers do not walk away in regular, predictable leaps. One possible explanation, put forward by Gould but with deep reservations and some doubt, is that bees are very smart animals who know what the biologist is up to and can imagine where he will turn up next with his sugar. Another possibility favored by Gould is that we simply do not understand the matter and need to learn more about bees. I like this answer, and it is my reason for putting the bee question at the top of my list.

My second question, addressed at large to the world of biology, concerns music. Surely, music (along with ordinary language) is as profound a problem for human biology as can be thought of, and I would like to see something done about it. A few years ago the German government set a large advisory committee to work on the question of what the next Max Planck Institute should be taking on as its scientific mission. The committee worked for a very long time and emerged with the recommendation that the new Max Planck Institute should be dedicated to the problem of music—what music is, why it is indispensable for human existence, what music really means—hard questions like that. The government, in its wisdom, turned down the idea, muttering something in administrative language about relevance, and there the matter rests. I shall take it as a sign of growing-up in the United States when we can assemble a similar committee for the same purpose and have the idea of the National Institute of Music approved and funded. I will not wait up for this to happen, but I can imagine it starting on a very small scale and with a very limited mission and a modest budget: a narrow question, like Why is *The Art of Fugue* so important and what does this single piece of music do to the human mind? Later on, there will be other questions, harder to deal with.

And while you are on your feet, Science, I have one last question, this time one closer to medicine. Some years ago,

Dr. Harold Wolff, professor of neurology at Cornell, conducted the following experiment. He hypnotized some healthy volunteer subjects, and while they were under deep hypnosis he touched their forearms with an ordinary pencil, which he told them was an extremely hot object; then he brought them out of the hypnotic state. In most cases, what happened was the prompt development of an area of redness and swelling at the skin site touched by the pencil, and in some subjects this went on to form a typical blister. I want to know all about that phenomenon. I also want to know how it happens that patients with intractable warts of long standing can have their warts instructed to drop off while under hypnosis.

Come to think of it, I would rather have a clear understanding of this phenomenon than anything else I can think of at the moment. The bees and the music can wait. If it is true, as it seems to be, that the human central nervous system can figure out how to go about creating a blister at a particular skin site, all on its own, or how to instruct its blood vessels, lymphocytes, and heaven knows what other participants in the tissues to eliminate a wart, then it is clear that the human nervous system has already evolved a vast distance beyond biomedical science. If I had a good wart I'd be happy to be a participant in this experiment, and I'll be glad any day to try my brain on a blister, but my motive for doing so would be less than worthy. If it worked I would feel gratified by the skill, excessively vain, and ready to dine out forever on the news that my own mind is so much smarter than I am.

Basic Science and the Pentagon

Basic science can be defined in a number of ways, but it is generally agreed that a central feature of the endeavor, which distinguishes it from other kinds of scholarship, is the absence of any predictable, usable product. In biology it has been immensely productive despite this apparent restriction; the discovery of mechanisms in nature has sometimes led, indirectly and often inadvertently, to methods for intervening in the same mechanisms. The control of infectious disease is only one example of the process at work.

In the last few years basic science has fallen on hard times in biology and medicine, even more so in physics and chemistry. Cosmology, perhaps the most basic of the sciences, is in the deepest trouble of all: the opportunities to pursue the exploration of our solar system, brilliantly begun, are being set aside because of the money shortage.

For the time being we seem to be stuck in a period of history when fundamental inquiry into matters of pure interest is being put off to the future. Most of what is going on in research today is aimed at clearly visible targets. It is applied science, intended to produce marketable products, capitalizing on the stores of basic knowledge that have accumulated thus far in a richly productive century.

I do not intend to quarrel here with this drift of things. If the nation has decided, through its elected leadership, to press ahead for practical applications and new technologies rather than to invest in pure inquiry for its own sake, I shall not argue here against that decision. I do not agree with it, I believe it to be unwise, but there it is. We live in a democracy.

But if that is to be the decision, I think it should be made fairly, cleanly, and unequivocally. This means that we should be scrutinizing all aspects of the national research effort in order to make sure that basic science *as a whole* is damped down. There should not be any exceptions.

Now, I will add to my definition of basic research in order to make clear the sort of science to be looked out for. It is not only the absence of a visible target and the lack of any imaginable product. It is the kind of research that depends on pure hunch, with nothing more than guesswork for the construction of the hypothesis to be tested. It is research carried out in an atmosphere of high uncertainty. The questions to be asked are in the nature of "What if?" questions, not the "How to?" questions that drive applied science along its even paths. It can be regarded as a dreamy sort of work, done by intelligent but highly impractical people, residents of an ivory tower, shielded from any contact with the realities of everyday human existence. Finally, it costs money, taxpayers' money, sometimes in very large amounts.

I wish I could report that research of this kind was being uniformly restrained, in accord with the perceived national policy, with equity all around for all fields of science. It is not so.

There is one large exception, an anomaly so enormous that it makes the whole policy look ridiculous.

The country, without seeming to notice that it is doing so, persists in one single venture of absolutely pure basic research, more basic in terms of the definitions described above than any other piece of research ever tried in all history. Moreover, the amount of public money being expended runs into many billions of dollars, enough to sustain all other fields of basic science for generations ahead, including the wildest imaginings of the astronomers and astrophysicists. With that kind of money we could be building Scarsdales on Mars if we had a mind to. We could be gardening out in the galaxy. We could free ourselves, our animals, and all our vegetation from disease. We could solve our energy problems and learn how to clean up after ourselves on our own suburban planet. We could begin paying attention to all our children, everywhere on the globe, and their children still to come. We could even begin learning enough about each other to begin growing up as a species, liking each other, on the way to loving each other.

The huge exception, the field of basic science that has been overlooked in all the cutbacks of funds, the area of fundamental inquiry that nobody seems to have noticed, is research on thermonuclear weapons.

I claim that this has to be classified as basic science on every count. It conforms to every item in any definition.

There is, to begin with, no usable product. To be sure, there are all those missiles, tucked away in their silos or riding through the underseas, but who would call those things a product? Who intends to use them, ever, for any purpose? Not us, we say. Not the Russians, they say. Not the Europeans, surely.

Moreover, an immense scientific establishment exists worldwide, with no research intention other than to make more of them, bigger and better, with more accurate systems for aiming them and guiding them to create new suns at whatever small spot on earth they choose. Worse, the new research programs to be added on, now that these nonproducts are in hand, are designed to protect this or that small spot against the other side's missiles. And underlying all the scientific questions is a deep, scientifically imponderable, central question. It is the paradigmatic "What if?" question of basic research: What if those things go off? Other imponderable questions: How do you protect a society against destruction if you have an *almost-perfect* antiballistic missile defense, one that will pick off with unfailing accuracy 950 out of 1000 missiles coming our way but will miss the other 50? Or if you have a system with unbelievable certainty, likely to miss only 10 of the 1000? What happens, then, when each missile is a ten- or twenty-megaton bomb, capable of vaporizing away whatever medium-sized city it happens to touch?

The present administration has no special fondness for the social and behavioral sciences, and the National Science Foundation is sharply reducing its funding—never generous at best—for these stepchildren of scholarship. Very well, the country will survive, and the disciplines of psychology, sociology, economics, and their siblings will have to eat grass until their time comes again. But the basic research enterprise involved in thermonuclear warfare contains a staggering array of behavioral research questions, the purest kind of social science, questions never before asked about human behavior, deep ambiguities approachable only in an atmosphere of almost total uncertainty. Should the country be providing the funding for these basic problems (I assume that the military must have the problems somewhere on its mind) if, at the same time, other fields in social science are to be put off? It is an unfairness, even a betrayal of principle.

But perhaps *I* am being unfair. Maybe these matters are not scheduled for study and do not exist in any line of the Pentagon budget. But then what? Who will be bringing in the data telling us what to expect when, say, five million of us vanish in twenty minutes and another five million are left behind with bone marrows burned out and skins in shreds, looking at what is left of the dead and waiting to die? Or, to magnify the problem to what will more likely be its true dimension, what will the few million survivors say to each other, or do to each other, at the moment when the other hundred millions are being transmuted back to the old interstellar dust? This, it seems to me, requires study, mandates study. Will no one be casting an anthropological eye at the dilemma to be faced when human beings cease being human?

The fundamental problem is the weaponry itself. Never mind the social sciences, for now. They can wait. If everyone agrees, as everyone seems to, that the weapons will never be let fly, never be used, never even be allowed out of the hatchways of their silos, and if, at the same time, everyone agrees, as everyone seems to agree, that they are indispensable not only for our security but for the security of the U.S.S.R. as well, then we are in the presence of a really great paradox.

Then there is the aspect of uncertainty, one of the rock-solid hallmarks for identifying basic science at its most basic. The problems raised by the mere existence of MIRVs, even more the questions raised by their presence in upper-space trajectories, are basic enough to pop the eyes. If they are flying over the North Pole, as they probably will, how good are the guidance data? Will the things wobble, yaw, shift course from Manhattan to the Sargasso Sea, to Yap, to wherever? Can we count on anything?

The dreamy, heavy-lidded, ivory-tower scientists at work on the weapons are also at work on nuclear defense, with all sorts of possibilities on their minds: laser beams to zap the intruding missiles while they are still safely aloft, anti-hydrogen-bomb hydrogen bombs, cities shielded by God knows what stealth armor, whole populations transported to safety under the hill, countries shielded by hope, by flags, by tears, by any old idea, by reassuring strings of words. Well, I claim this is basic research, and it should be stopped.

Or, if it is to be continued, and funded, I want in. As a citizen and a sometime scientist, I claim rights to a grant, part of the $200 billion or whatever it is. As it happens, I happen to have an idea, as good an idea as any of the thermonuclear notions I've heard about so far, worth a few billions. The United States now has around 300,000 American troops stationed in Europe, put there some years ago as a significant token of our determination to defend Europe. We assert, correctly I am sure, that if any nuclear bomb of any size down to the neatest, cleanest, most tactfully tactical "theater" bomb should ever be launched against those troops this country will surely let fly some of ours in return. We acknowledge, most of us, anyway, that that will lead to exchanges of the larger then still larger weapons, across the Atlantic, megaton after megaton, until everything is made dead.

To avoid this outcome (which in a *New York Times Magazine* article two military authorities, proponents of thermonuclear research, referred to in passing as a "dismal prospect"), I suggest that we use these hostages differently, and persuade the Russians to put up 300,000 troops as tokens of their good faith to be used symmetrically. Bring all the Russian troops to this country and put them on trains, different trains, different lines. Send all 300,000 Americans to the Soviet Union and put them on *their* trains, at random. Let the trains go off on their normal schedules. Once begun, the program will provide absolute assurance to each side that the other side can never know precisely where its 300,000 soldiers are at any given time. Hence, neither country will send off any missiles, for fear of killing a large number of its own people. American and Russian train schedules are both matters of public knowledge, well enough understood so that both sides can feel secure in their unpredictability.

It is, I grant, a little like the first plan for putting MX missiles on underground railways, but a real improvement, less expensive and far simpler than running those costly missiles around in Utah or Nevada. It would be the ultimate defense strategy. Enough money would be saved to provide real, old-fashioned Pullman accommodations for all the soldiers, good filling meals, marvelous views, movies, sleepers, the works. The entire railroad systems of both countries would be restored to solvency, maybe even enriched sufficiently to improve the roadbeds. Everyone would benefit, and no one would get hurt. And no one would fire from abroad, not a single missile, for fear of hitting them.

A side benefit, a spin-off, so to speak, might be the effect on the train-riding troops themselves. Looking out of the windows at the countryside views of both countries—the foreign scenery, the farms and gardens—they would catch glimpses, especially as their decoy trains entered the stations of various towns and cities, of something neither they nor their commanders may have realized before: there are people out there.

Science and "Science"

I remain puzzled over how to classify the science and technology underlying nuclear warfare. Because of the high degree of uncertainty involved in this sort of research, as well as the ambiguity and unpredictability of its outcome, I asserted earlier that it should be classed as entirely basic science, by definition, and should therefore be subject to the same budgetary constraints as all the rest of basic research in the nation.

Since then I have changed my mind. I recant, partly under the influence of the Defense Department's latest announced R & D plans for fighting a "protracted" nuclear war. I take it back. This kind of military research is not at all like any science I have ever heard of, basic or applied. It is a new, peculiar kind of endeavor for which some other term must be invented. It resembles, more closely, an endless game. I suppose you could argue that scientific research is also a sort of game, but there is a difference: ordinary games finish at some point; there are winners and losers. Neither of these seems possible under the rules of nuclear "science," and the game seems designed to go on forever. I have placed the word in quotation marks, provisionally, until I can think of a suitable euphemism.

There are many differences, but one that is decisive and compelling. In science in general, one characteristic feature is the awareness of error in the selection and pursuit of a problem. This is the most commonplace of criteria: if a scientist is going to engage in research of any kind, he has to have it on his mind, from the outset, that he may be on to a dud. You can tell a world-class scientist from the run-of-the-mill investigator by the speed with which he recognizes that he is heading into a blind alley. Blind alleys and garden paths leading nowhere are the principal hazards in research.

Everyone in science knows about this, although it is not much talked about. Most scientific hypotheses, including what seem the brightest and best, turn out to be wrong. I would guess that the luckiest and most productive of investigators is right in his original notion, the guess with which he starts his work, about once out of a hundred tries, at his peak. What counts is his instinct for spotting wrongness, his willingness to give up on a favorite conviction, his readiness to quit and shift to a better project. Insoluble problems abound. It takes almost as much good judgment to recognize these when they turn up as to perceive quickly the ones that can be settled, solved, nailed down once and for all by research.

There are two ways of clouding the vision of a working scientist in making this discrimination (assuming that he is of average intellectual stature). One is money. If the funds for a particular research project are coming in over his head in cascades, the scientist may be misled into thinking that he is on to a good thing, no matter what his data show. I can only suspect this to be a possibility, having never seen it happen. Second, the researcher may be led down his garden path by his equipment. If he is in possession of sophisticated instruments of great power, and if he is being assured that whatever other new instruments he can think of will be delivered to the door of his laboratory tomorrow, he may find it difficult to stop himself on a dead road of inquiry, even if he knows it to be dead. I have long believed that there is no scientist alive whose career could not be terminated by an enemy, if the enemy were capable of increasing the laboratory's budget by tenfold or any-fold overnight and, as well, assuring access immediately to any instrument within reach of the victim's imagination.

Maybe one or the other or both of these things are responsible for what is going on in defense "science" today. Certainly, the cascades of money are out there, with more promised to come, enough to convince any individual researcher that the project must be a good one even if he doesn't understand why. And meanwhile, like enormous, irresistible, gleaming and spinning toys, there are the missiles and their warheads, each one more destructive than one thousand Hiroshima bombs, loaded with magnificent navigational equipment more fun to play with than anything else on earth or in space. Any technological whim can be promised instant gratification on demand. Toys indeed.

Still, the damned things are not going to work and the

"science" is leading nowhere. Everyone who ought to know knows this, and almost everyone agrees, even in public. It is endgame, a dud, a piece of nonscience flawed enough to qualify as nonsense. There is no way to design or redesign these weapons so that they can ever be used to win a war or even to fight a war, and no technological fix within the grasp of human imagination that can assure defense against them—if what is to be defended is human society.

And yet the "science" goes on, one side adding an embellishment to its threat to devastate the cities of the other side, the other side then dangling a matching ornament; one side donning a new horrifying mask, the other waving a scary banner; a war dance on one side, a counter-dance on the other.

Lord Solly Zuckerman, a longtime science adviser to the British government on science in general and military science and technology in particular, has written a book on this, the best book on the problem that I have read in recent years: *Nuclear Illusion and Reality.* Zuckerman believes that the "technicians" are the main source of our trouble today. It is interesting that he uses the term as reservedly as do many proper scientists, even when, as in the case of physicists, they are talking about themselves. They do not refer to the workers in the field as scientists or engineers; they are the "technicians."

Well, Lord Zuckerman says, the technicians have been far too long in charge of the world's destiny, if you accept that the world's destiny will, in part, anyway, be determined by how we all come out in nuclear warfare. He maintains, as do most science advisers in most nations, that "nuclear weapons and nuclear weapons systems are not weapons of war." He says, rather more bluntly than most scientists seem willing to say, that the scientific community itself must carry a heavy burden of responsibility for today's dilemma. "The scientists who work in the defense departments of governments, or in defense industries, are not apostles of peace. Political and military leaders should cease seeking shelter behind the backs of those 'experts' who take what is usually called the harder line. . . . If the bridge itself is not to become submerged, the politicians will have to take charge of the technical men." There it is again, a scientist writing of the work of other scientists, now obliged to refer to them as "technical men."

It does not often happen that scientists come out with public declarations that an avenue of scientific research should be blocked by public action. Most of the time, the entire scientific community maintains that deep research is unpredictable because it is aimed at uncovering brand-new information about nature. To stop science at any point, in any field, risks a restriction on human knowledge, and scientists as a group believe in their hearts that knowledge and understanding of the world are what the world most needs. There have, of course, been a few exceptions: many eminent physicists and mathematicians were deeply opposed, back in the late 1940s, to any research that might lead to the development of the hydrogen bomb; more recently, a smaller group of biologists advised delay in the pursuit of recombinant-DNA techniques until the safety of the method could be assured.

But it is possible to oppose the kind of research involved in nuclear warfare on grounds that have nothing to do with the traditional openness and curiosity of the scientific mind. This is not an instance of scientists blocking science itself. What is to be blocked, if it can indeed be blocked, is not real science. It has a single objective that cannot conceivably be attained: national security, which lies as far beyond its reach as transmutation for the early alchemists. Indeed, as Zuckerman points out in his book, "the continued growth of nuclear arsenals not only fails to increase, but actually decreases, national security."

Zuckerman puts it in the coldest language. The development of nuclear weapons, he says, derives its momentum not from any notions of military science or national policy on either side but "from the minds of enthusiastic technicians plying their trade in the weapons laboratories." He is not talking about people he would be willing to call scientists.

It is hard to see how the nations possessing nuclear weaponry can come to an agreement on arms reduction by simply throwing away part of the arsenals now on hand, although this must somehow be accomplished, sooner or later. But it is even more important, in the interest of long-term stability, that all research efforts to devise better weapons and better defense systems be ended. The world needs a multilateral treaty under which all parties will agree to stop the flow of money into their nuclear R & D laboratories, to stop all testing of weapons systems, and, above all, to stop the "technicians" from cooking up new ideas.

On the Need for Asylums

From time to time, medical science has achieved an indisputable triumph that is pure benefit for all levels of society and deserving of such terms as "breakthrough" and "medical miracle." It is not a long list, but the items are solid bits of encouragement for the future. The conquests of tuberculosis, smallpox, and syphilis of the central nervous system should be at the top of anyone's list. Rheumatic fever, the most common cause of heart disease forty years ago, has become a rare, almost exotic disorder, thanks to the introduction of antibiotics for treating streptococcal sore throat. Some forms of cancer—notably childhood leukemias, Hodgkin's disease, and certain sarcomas affecting young people—have become curable in a high proportion of patients. Poliomyelitis is no longer with us.

But there is still a formidable agenda of diseases for which there are no cures, needing much more research before their underlying mechanisms can be brought to light. Among these diseases are some for which we have only half-way technologies to offer, measures that turn out to be much more costly than we had guessed and only partly, sometimes marginally, effective. The transplantation of major organs has become successful, but only for a relatively small number of patients with damaged kidneys and hearts, and at a financial cost much too high for applying the technologies on a wide scale. Very large numbers of patients with these fatal illnesses have no access to such treatments. Renal dialysis makes it possible to live for many months, even a few years, with failed kidneys, but it is a hard life.

The overestimation of the value of an advance in medicine can lead to more trouble than anyone can foresee, and a lot of careful thought and analysis ought to be invested before any technology is turned loose on the marketplace. It begins to look as if coronary bypass surgery, for example, is an indispensable operation for a limited number of people, but it was probably not necessary for the large number in whom the expensive procedure has already been employed.

There are other examples of this sort of premature, sweeping adoption of new measures in medicine. Probably none has resulted in more untoward social damage than the unpredicted, indirect but calamitous effects of the widespread introduction twenty or so years ago of Thorazine and its chemical relatives for the treatment of schizophrenia. For a while, when it was first used in state hospitals for the insane, the new line of drugs seemed miraculous indeed. Patients whose hallucinations and delusions impelled them to wild, uncontrollable behavior were discovered to be so calmed by the treatment as to make possible the closing down of many of the locked wards in asylums. Patients with milder forms of schizophrenia could return, at least temporarily, to life outside the institutions. It was the first real advance in the treatment of severe mental disease, and the whole world of psychiatry seemed to have been transformed. Psychopharmacology became, overnight, a bright new discipline in medicine.

Then came the side effect. Not a medical side effect (although there were some of these) but a political one, and a disaster. On the assumption that the new drugs made hospitalization unnecessary, two social policies were launched with the enthusiastic agreement of both the professional psychiatric community and the governmental agencies responsible for the care of the mentally ill. Brand-new institutions, ambitiously designated "community mental health centers," were deployed across the country. These centers were to be the source of the new technology for treating schizophrenia, along with all other sorts of mental illness: in theory, patients would come to the clinics and be given the needed drugs, and, when necessary, psychotherapy. And at the same time orders came down that most of the patients living in the state hospitals be discharged forthwith to their homes or, lacking homes, to other quarters in the community.

For a while it looked like the best of worlds, on paper, anyway. Brochures with handsome charts were issued by state and federal agencies displaying the plummeting curves of state hospital occupancy, with the lines coinciding marvelously with the introduction of the new drugs. No one noted that the occupancy of private mental hospitals rose at the same time—though it could not rise very high, with the annual cost of such hospitalization running around $40,000 per bed. The term "breakthrough" was used over and over again, but after a little while it came to be something more like a breakout. The mentally ill were out of the hospital, but in many cases they were simply out on the streets, less agitated but lost, still disabled but now uncared for. The community mental health centers were not designed to take on the task of custodial care. They could serve as shelters only during the hours of appointment, not at night.

All this is still going on, and it is not working. To be sure, the drugs do work—but only to the extent of allaying some of the most distressing manifestations of schizophrenia. They do not turn the disease off. The evidences of the mind's unhinging are still there, coming and going in cycles of remission and exacerbation just as they have always done since schizophrenia was first described. Some patients recover spontaneously and for good, as some have always done. The chronically and permanently disabled are better off because they are in lesser degrees of mental torment when they have their medication; but they are at the same time much worse off because they can no longer find refuge when they are in need of it. They are, instead, out on the streets, or down in the subways, or wandering in the parks, or confined in shabby rooms in the shabbiest hotels, alone. Or perhaps they are living at home, but not many of them living happily; nor are many of their families happy to have them at home. One of the high risks of severe mental disease is suicide, and many of these abandoned patients choose this way out, with no one to stop them. It is an appalling situation.

It is claimed that the old state hospitals were even more appalling. They were called warehouses for the insane, incapable of curing anything, more likely to make it worse by the process known in psychiatric circles as "institutionalization," a steady downhill course toward total dependency on the very bleakness of the institution itself. The places were badly managed, always understaffed, repellent to doctors, nurses, and all the other people needed for the care of those with sick minds. Better off without them, it was said. Cheaper too, although this wasn't said so openly.

What never seems to have been thought of, or at least never discussed publicly, was changing the state hospitals from bad to good institutions, given the opportunity for vastly improved care that came along with the drugs. It was partly the history of such places that got in the way. For centuries the madhouses, as they were called, served no purpose beyond keeping deranged people out of the public view. Despite efforts at reform in the late nineteenth and early twentieth centuries, they remained essentially lockups.

But now it is becoming plain that life in the state hospitals, bad as it was, was better than life in the subways or in the doorways of downtown streets, late on cold nights with nothing in the shopping bag to keep a body warm, and no protection at all against molestation by predators or the sudden urge for self-destruction. What now?

We should restore the state hospital system, improve it, expand it if necessary, and spend enough money to ensure that the patients who must live in these institutions will be able to come in off the streets and live in decency and warmth, under the care of adequately paid, competent professionals and compassionate surrogate friends.

If there is not enough money, there are ways to save. There was a time when many doctors were glad to volunteer their services on a part-time basis, indeed competed to do so, unpaid by state or federal funds and unreimbursed by insurance companies, in order to look after people unable to care for themselves. We should be looking around again for such doctors, not necessarily specialists in psychiatric medicine, but well-trained physicians possessing affection for people in trouble—a quality on which recruitment to the profession of medicine has always, we hope, been based. We cannot leave the situation of insane human beings where it is today.

A society can be judged by the way it treats its most disadvantaged, its least beloved, its mad. As things now stand, we must be judged a poor lot, and it is time to mend our ways.

Altruism

Altruism has always been one of biology's deep mysteries. Why should any animal, off on its own, specified and labeled by all sorts of signals as its individual self, choose to give up its life in aid of someone else? Nature, long viewed as a wild, chaotic battlefield swarmed across by more than ten million different species, comprising unnumbered billions of competing selves locked in endless combat, offers only one sure measure of success: survival. Survival, in the cool economics of biology, means simply the persistence of one's own genes in the generations to follow.

At first glance, it seems an unnatural act, a violation of nature, to give away one's life, or even one's possessions, to another. And yet, in the face of improbability, examples of altruism abound. When a worker bee, patrolling the frontiers of the hive, senses the nearness of a human intruder, the bee's attack is pure, unqualified suicide; the sting is barbed, and in the act of pulling away the insect is fatally injured. Other varieties of social insects, most spectacularly the ants and higher termites, contain castes of soldiers for whom self-sacrifice is an everyday chore.

It is easy to dismiss the problem by saying that "altruism" is the wrong technical term for behavior of this kind. The word is a human word, pieced together to describe an unusual aspect of human behavior, and we should not be using it for the behavior of mindless automata. A honeybee has no connection to creatures like us, no brain for figuring out the future, no way of predicting the inevitable outcome of that sting.

But the meditation of the 50,000 or so connected minds of a whole hive is not so easy to dismiss. A multitude of bees can tell the time of day, calculate the geometry of the sun's position, argue about the best location for the next swarm. Bees do a lot of close observing of other bees; maybe they know what follows stinging and do it anyway.

Altruism is not restricted to the social insects, in any case. Birds risk their lives, sometimes lose them, in efforts to distract the attention of predators from the nest. Among baboons, zebras, moose, wildebeests, and wild dogs there are always stubbornly fated guardians, prepared to be done in first in order to buy time for the herd to escape.

It is genetically determined behavior, no doubt about it. Animals have genes for altruism, and those genes have been selected in the evolution of many creatures because of the advantage they confer for the continuing survival of the species. It is, looked at in this way, not the emotion-laden problem that we feel when we try to put ourselves in the animal's place; it is just another plain fact of life, perhaps not as hard a fact as some others, something rather nice, in fact, to think about.

J. B. S. Haldane, the eminent British geneticist, summarized the chilly arithmetic of the problem by announcing, "I would give up my life for two brothers or eight cousins." This calculates the requirement for ultimate self-interest: the preservation and survival of an individual's complement of genes. Trivers, Hamilton, and others have constructed mathematical models to account nicely for the altruistic behavior of social insects, quantifying the self-serving profit for the genes of the defending bee in the act of tearing its abdomen apart. The hive is filled with siblings, ready to carry the *persona* of the dying bee through all the hive's succeeding generations. Altruism is based on kinship; by preserving kin, one preserves one's self. In a sense.

Haldane's prediction has the sound of a beginning sequence: two brothers, eight (presumably) first cousins, and then another series of much larger numbers of more distant relatives. Where does the influence tail off? At what point does the sharing of the putative altruist's genes become so diluted as to be meaningless? Would the line on a graph charting altruism plummet to zero soon after those eight cousins, or is it a long, gradual slope? When the combat marine throws himself belly-down on the live grenade in order to preserve the rest of his platoon, is this the same sort of altruism, or is this an act without any technically biological meaning? Surely the marine's genes, most of them, will be blown away forever; the statistical likelihood of having two brothers or eight cousins in that platoon is extremely small. And yet there he is, belly-down as if by instinct, and the same kind of event has been recorded often enough in wartime to make it seem a natural human act, normal enough, even though rare, to warrant the stocking of medals by the armed services.

At what point do our genetic ties to each other become so remote that we feel no instinctual urge to help? I can imagine an argument about this, with two sides, but it would be a highly speculative discussion, not by any means pointless but still impossible to settle one way or the other. One side might assert, with total justification, that altruistic behavior among human beings has nothing at all to do with genetics, that there is no such thing as a gene for self-sacrifice, not even a gene for helpfulness, or concern, or even affection. These are attributes that must be learned from society, acquired by cultures, taught by example. The other side could maintain, with equal justification, since the facts are not known, precisely the opposite position: we get along together in human society because we are genetically designed to be social animals, and we are obliged, by instructions from our genes, to be useful to each other. This side would argue further that when we behave badly, killing or maiming or snatching, we are acting on misleading information learned from the wrong kinds of society we put together; if our cultures were not deformed, we would be better company, paying attention to what our genes are telling us.

For the purposes of the moment I shall take the side of the sociobiologists because I wish to carry their side of the argument a certain distance afield, beyond the human realm. I have no difficulty in imagining a close enough resemblance among the genomes of all human beings, of all races and geographic origins, to warrant a biological mandate for all of us to do whatever we can to keep the rest of us, the species, alive. I maintain, despite the moment's evidence against the claim, that we are born and grow up with a fondness for each other, and we have genes for that. We can be talked out of it, for the genetic message is like a distant music and some of us are hard-of-hearing. Societies are noisy affairs, drowning out the sound of ourselves and our connection. Hard-of-hearing, we go to war. Stone-deaf, we make thermonuclear missiles. Nonetheless, the music is there, waiting for more listeners.

But the matter does not end with our species. If we are to take seriously the notion that the sharing of similar genes imposes a responsibility on the sharers to sustain each other, and if I am right in guessing that even very distant cousins carry at least traces of this responsibility and will act on it whenever they can, then the whole world becomes something to be concerned about on solidly scientific, reductionist, genetic grounds. For we have cousins more than we can count, and they are all over the place, run by genes so similar to ours that the differences are minor technicalities. All of us, men, women, children, fish, sea grass, sandworms, dolphins, hamsters, and soil bacteria, everything alive on the planet, roll ourselves along through all our generations by replicating DNA and RNA, and although the alignments of nucleotides within these molecules are different in different species, the molecules themselves are fundamentally the same substance. We make our proteins in the same old way, and many of the enzymes most needed for cellular life are everywhere identical.

This is, in fact, the way it should be. If cousins are defined by common descent, the human family is only one small and very recent addition to a much larger family in a tree extending back at least 3.5 billion years. Our common ancestor was a single cell from which all subsequent cells derived, most likely a cell resembling one of today's bacteria in today's soil. For almost three-fourths of the earth's life, cells of that first kind were the whole biosphere. It was less than a billion years ago that cells like ours appeared in the first marine invertebrates, and these were somehow pieced together by the joining up and fusion of the earlier primitive cells, retaining the same blood lines. Some of the joiners, bacteria that had learned how to use oxygen, are with us still, part of our flesh, lodged inside the cells of all animals, all plants, moving us from place to place and doing our breathing for us. Now there's a set of cousins!

Even if I try to discount the other genetic similarities linking human beings to all other creatures by common descent, the existence of these beings in my cells is enough, in itself, to relate me to the chestnut tree in my backyard and to the squirrel in that tree.

There ought to be a mathematics for connections like this before claiming any kinship function, but the numbers are too big. At the same time, even if we wanted to, we cannot think the sense of obligation away. It is there, maybe in our genes for the recognition of cousins, or, if not, it ought to be there in our intellects for having learned about the matter. Altruism, in its biological sense, is required of us. We have an enormous family to look after, or perhaps that assumes too much, making us sound like official gardeners and zookeepers for the planet, responsibilities for which we are probably not yet grown-up enough. We may need new technical terms for concern, respect, affection, substitutes for altruism. But at least we should acknowledge the family ties and, with them, the obligations. If we do it wrong, scattering pollutants, clouding the atmosphere with too much carbon dioxide, extinguishing the thin carapace of ozone, burning up the forests, dropping the bombs, rampaging at large through nature as though we owned the place, there will be a lot of paying back to do and, at the end, nothing to pay back with.

Falsity and Failure

Two friends of mine, eminent scientists with high responsibilities for science management and policy, were recently called before a congressional committee to testify in defense of the morals of American research. Why is it, they were asked, that there have been so many instances of outright fraud and plagiarism in recent years, so many publications of experiments never actually performed, so much fudging of data?

At about the same time, articles about falsification of research—particularly in biomedical science—appeared in *The New York Times* and in two respected and widely read technical periodicals, *Nature* (published in London) and *Science* (in Washington). The general drift of the thoughtful, worried essays was that the reported instances of deliberate mistruth on the part of scientists seem to be on the increase, and the self-monitoring system, traditionally relied upon to spot and immediately expose all cases of faked data, appears to be malfunctioning.

The list of fraudulent research reports is not a long one, but several of the studies were carried out within the walls of the country's most distinguished scientific institutions, long regarded as models of scientific probity. *Science* stated flat out, in its April 10, 1981, issue: "There is little doubt that a dark side of science has emerged during the past decade. . . . Four major cases of cheating in biomedical research came to light in 1980 alone with some observers in the lay press calling it a 'crime wave.'" It is the same list in all the reports: the case of a pathologist knowingly employing a contaminated cell-culture line, two junior researchers who plagiarized work already done by others, a clinical investigator found to have inserted bogus data on cancer chemotherapy into the project's computer. None of these studies involved crucial issues of science; the papers in question dealt with relatively minor matters, unlikely to upheave any field but requiring, nonetheless, a significant waste of time and money in other laboratories attempting to confirm the unconfirmable. The real damage has been done to the public confidence in the scientific method, and there are apprehensions within the scientific community itself that someone, somewhere, perhaps in Washington, will begin framing new regulations to ensure exactitude and honesty in an endeavor that has always prided itself, and depended for its very progress, on these two characteristics.

Now that the issue has surfaced so publicly it is likely that story will lead to story, and there will be more speculations and skepticism about any piece of science that seems to carry surprising or unorthodox implications. Indeed, a number of old stories are being exhumed and revived, as though to reveal a pattern of habitual falsehood in the process of science: Ptolemy and his unearthly second-century A.D. data establishing the sun's movement around the earth, supposed examples of seventeenth-century fudging in Newton's calculations, even some small questions about the perfection of Gregor Mendel's classical (and absolutely solid) generalizations about plant genetics one hundred years ago. Lumped in with these are some outright examples of bent science: Cyril Burt's 1930s data on the inheritance of intelligence in identical twins, the falsified synthesis of a cellular protein by a postdoctoral student at the Rockefeller Institute twenty years ago, and the notorious episode of skin-graft fabrication at Sloan-Kettering in 1974. These can, if you like, be made to seem all of a piece, a constantly spreading blot on the record of science. Or, if you prefer (and I do prefer), they can be viewed as anomalies, the work of researchers with unhinged minds or, as in the cases of Newton and Mendel, gross exaggerations of the fallibility of even superb scientists.

It is an impossibility for a scientist to fake his results and get away with it, unless he is lucky enough to have the faked data conform, in every fine detail, to a guessed-at truth about nature (the probability of this kind of luck is exceedingly small), or unless the work he describes is too trivial to be of interest to other investigators. Either way, he cannot win. If he reports something of genuine significance, he knows for a certainty that other workers will repeat his experiments, or try to, and if he has cooked his data the word will soon be out, to the ruin of his career. If he has plagiarized someone else's paper, the computer retrieval systems available to scientific libraries everywhere will catch him at it, sooner or later, with the same result.

In short, the system does indeed work, and the fact that only four instances of scientific malfeasance have been identified in a year during which some 18,000 research projects were sponsored by the National Institutes of Health means just what the fact says: such events are extremely rare. This is not a claim that scientists are necessarily, by nature, an impeccably honest lot, although I am convinced that all the best ones are. It says, simply, that people are not inclined to try cheating in a game where cheating leads almost inevitably to losing.

You have only to glance through other pages of the same issue of *Nature* in which the editorial comment on fraud appears to catch a sense of how the system really works, and works at its very best. There are two extensive papers dealing with an important and fascinating question raised last year by a Canadian group of immunologists, who had found evidence suggesting, cautiously but not conclusively, that mice can inherit through the male line an acquired abnormality in their immune cells known technically as "tolerance." If true, the claim would support nothing less than Lamarckianism, long since jettisoned from biology; it would be an upheaval indeed to face again, as an open problem, the question of the inheritance of acquired characteristics. The two new papers explore the matter in elegantly designed and meticulously executed experiments, with the conclusion that the Canadian work could not be confirmed. There is, in this instance, no question at all of contrived data or even the misguided reading of results; it is a typical instance of disagreement in science, something that happens whenever major ideas are under exploration. It will, in this case, lead to more work in the three laboratories now caught up in the problem and no doubt in others not yet involved. It may also lead to new knowledge, a deeper comprehension of immunology, and conceivably to something surprising, even if the Canadians turn out at the end to have been totally wrong.

In the same journal there are three marvelous papers—one from the University of Cambridge, two from Cal Tech—that will cause even more of a stir in biomedical science, generating surprise, argument, and new bursts of research in laboratories around the world. The genetic composition of human mitochondria has been elucidated, and these structures, long believed to be the descendants of bacteria living as permanent lodgers inside all nucleated cells, are turning out to have an arrangement of their genes like nothing else on earth: they display an astonishing economy in their circles of DNA, and in some respects the genetic code is different from what has, up to now, been regarded as a universal code. It is something quite new, unorthodox, unexpected, and therefore certain to be challenged but also likely to be repeated, confirmed, and extended. Molecular genetics may then be moved on to new ground, new explanations for the origin of mitochondria will be thought up and tested, and science itself will be off and running in a new direction.

With work like this going on in the pages of a single issue of *Nature*, and with similar things to be read in *Science*, week after week these days, I cannot find time to worry so much about falsity and fraud. Only to reflect that my dictionary gives the Latin root for "falsity" as *fallere*, which is the same root for the word "failure."

On Medicine and the Bomb

In the complicated but steadily illuminating and linked fields of immunology, genetics, and cancer research, it has become a routine technical maneuver to transplant the bone-marrow cells of one mouse to a mouse of a different line. This can be accomplished by irradiating the recipient mouse with a lethal dose of X rays, enough to destroy all the immune cells and their progenitors, and replacing them with the donor's marrow cells. If the new cells are close enough in their genetic labels to the recipient's own body cells, the marrow will flourish and the mouse will live out a normal life span. Of course, if the donor cells are not closely matched, they will recognize the difference between themselves and the recipient's tissues, and the result, the so-called graft-versus-host reaction, will kill the recipient in the same way that a skin graft from a foreign mouse is destroyed by the lymphocytes of a recipient.

It is a neat biological trick, made possible by detailed knowledge of the genetics involved in graft rejection. Any new bone-marrow cells can survive and repopulate the recipient's defense apparatus provided the markers on the cell surfaces are the same as those of the donor, and precise techniques are now available for identifying these markers in advance.

Something like this can be done in human beings, and the technique of bone-marrow transplantation is now becoming available for patients whose marrows are deficient for one reason or another. It is especially useful in the treatment of leukemia, where the elimination of leukemic cells by X ray and chemotherapy sometimes causes the simultaneous destruction of the patient's own immune cells, which must then be replaced if the patient is to survive. It is a formidable procedure, requiring the availability of tissue-match donors (usually members of the patient's family), and involving extremely expensive and highly specialized physical facilities—rooms equipped for absolute sterility to prevent infection while the new cells are beginning to propagate. Not many hospitals are outfitted with units for this kind of work, perhaps twenty or twenty-five in the United States, and each of them can take on only a few patients at a time. The doctors and nurses who work in such units are among the most specialized of clinical professionals, and there are not many of them. All in all, it is an enormously costly venture, feasible in only a few places but justifiable by the real prospect of new knowledge from the associated research going on in each unit, and of course by the lifesaving nature of the procedure when it works.

This, then, is the scale on which contemporary medicine

possesses a technology for the treatment of lethal X-irradiation.

The therapy of burns has improved considerably in recent years. Patients with extensively burned skin who would have died ten years ago are now, from time to time, being saved from death. The hospital facilities needed for this accomplishment are comparable, in their technical complexity and cost, to the units used for bone-marrow transplantation. Isolation rooms with special atmospheric controls to eliminate all microbes from the air are needed, plus teams of trained professionals to oversee all the countless details of management. It is still a discouraging undertaking, requiring doctors and nurses of high spirit and determination, but it works often enough to warrant the installation of such units in a limited number of medical centers. Some of these places can handle as many as thirty or forty patients at a time, but no more than that number.

The surgical treatment of overwhelming trauma underwent a technological transformation during the Korean and Vietnam wars, and it is now possible to do all sorts of things to save the lives of injured people—arteries and nerves can be successfully reconnected, severed limbs sewn back in place, blood substitutes infused, shock prevented, massive damage to internal organs repaired. Here also, special units with highly trained people are essential, elaborate facilities for rapid transport to the hospital are crucial, and the number of patients that can be handled by a unit is minimal.

These are genuine advances in medical science. The medical profession can be proud of them, and the public can be confident that work of this kind will steadily improve in the future. The prospects for surviving various kinds of injury that used to be uniformly fatal are better now than at any other time in history.

If there were enough money, these things could be scaled up to meet the country's normal everyday needs with tailor-made centers for the treatment of radiation injury, burns, and massive trauma spotted here and there in all major urban centers, linked to outlying areas by helicopter ambulances. It would cost large sums to build and maintain, but the scores, maybe hundreds, of lives saved would warrant the cost.

The Department of Defense ought to have a vested interest in enhancing this array of technologies, and I suppose it does. I take it for granted that substantial sums are being spent from the R & D funds of that agency to improve matters still further. In any conventional war, the capacity to rescue injured personnel from death on the battlefield does more than simply restore manpower to the lines: its effect on troop morale has traditionally been incalculable.

But I wonder if the hearts of the long-range planners in DOD can really be in it.

Military budgets have to be put together with the same analytic scrutiny of potential costs versus benefits that underlies the construction of civilian budgets, allowing for the necessarily different meanings assigned by the military to the terms "cost" and "benefit." It is at least agreed that money should not be spent on things that will turn out to be of no use at all. The people in the Pentagon offices and their counterparts in the Kremlin where the questions of coping with war injuries are dealt with must be having a hard time of it these days, looking ahead as they must look to the possibility of thermonuclear war. Any sensible analyst in such an office would be tempted to scratch off all the expense items related to surgical care of the irradiated, burned, and blasted, the men, women, and children with empty bone marrows and vaporized skin. What conceivable benefit can come from sinking money in hospitals subject to instant combustion, only capable of salvaging, at their intact best, a few hundred of the victims who will be lying out there in the hundreds of thousands? There exists no medical technology that can cope with the certain outcome of just one small, neat, so-called tactical bomb exploded over a battlefield. As for the problem raised by a single large bomb, say a twenty-megaton missile (equivalent to approximately two thousand Hiroshimas) dropped on New York City or Moscow, with the dead and dying in the millions, what would medical technology be good for? As the saying goes, forget it. Think of something else. Get a computer running somewhere in a cave, to estimate the likely numbers of the lucky dead.

The doctors of the world know about this, of course. They have known about it since the 1945 Hiroshima and Nagasaki "episodes," but it has dawned on them only in the last few years that the public at large may not understand. Some of the physicians in this country and abroad are forming new organizations for the declared purpose of making it plain to everyone that modern medicine has nothing whatever to offer, not even a token benefit, in the event of thermonuclear war. Unlike their response to other conceivable disasters, they do not talk of the need for more research or ask for more money to expand existing facilities. What they say is, in effect, count us out.

It is not a problem that has any real connection to politics. Doctors are not necessarily pacifists, and they come in all sorts of ideological stripes. What they have on their minds and should be trying to tell the world, in the hope that their collective professional opinion will gain public attention and perhaps catch the ears of political and military leaders everywhere, is simply this: if you go ahead with this business, the casualties you will instantly produce are beyond the reach of any health-care system. Since such systems here and abroad are based in urban centers, they will vanish in the first artificial suns, but even if they were miraculously to survive they could make no difference, not even a marginal difference.

I wish the psychiatrists and social scientists were further along in their fields than they seem to be. We need, in a hurry, some professionals who can tell us what has gone wrong in the minds of statesmen in this generation. How is it possible for so many people with the outward appearance of steadiness and authority, intelligent and convincing enough to have reached the highest positions in the governments of the world, to have lost so completely their sense of responsibility for the human beings to whom they are accountable? Their obsession with stockpiling nuclear armaments and their urgency in laying out detailed plans for using them have, at the core, aspects of what we would be calling craziness in other people, under other circumstances. Just before they let fly everything at their disposal, and this uniquely intelligent species begins to go down, it would be a small comfort to understand how it happened to happen. Our descendants, if there are any, will surely want to know.

The Problem of Dementia

I have always tended to agree with those who criticize the government for legislating special funds for individual diseases (the "disease-of-the-month" syndrome, as it is termed) at cost to the country's broader and, on balance, more productive programs in basic, undifferentiated science. The evidence is convincing enough: we have been learning more ways into the center of human disease mechanisms by studying normal biological processes than by mounting frontal, targeted assaults on one disease after another. When, for example, we have found our way to the epicenter of the cancer problem, as sooner or later we shall, armed with enough deep information to switch a neoplastic cell back to the normal mode of life, it will be because the crucial information will have emerged in its own time from basic research on normal cells, most of it likely to be coming from applications of the recombinant-DNA technique to fundamental cell biology.

I would, however, make one exception to the nontargeting rule, and push for special consideration and high priority for one particular human disease, not a disease-of-the-month but a disease-of-the-century, the brain disease that afflicts increasing members of our population because of the increasing population of older people in the society—senility, or, as it is now termed, senile dementia. The major form of the disorder, Alzheimer's disease, affects more than 500,000 people over the age of fifty, most of them in their seventies and eighties. It is responsible for most of the beds in the country's nursing homes, at a cost exceeding $10 billion now and scheduled to rise to $40 billion or more within the next few years.

It is the worst of all diseases, not just for what it does to the patient, but for its devastating effects on families and friends. It begins with the loss of learned skills—arithmetic and typing, for instance—and progresses inexorably to a total shutting down of the mind. It is not in itself lethal, unmercifully; patients go on and on living, essentially brainless but otherwise healthy, into advanced age, unless lucky enough to be saved by pneumonia.

It is not, as we used to think, simply an aspect of aging or a natural part of the human condition, nor is it due to hardening of the arteries or anything else we know about. It remains an unsolved mystery.

One acceptable guess, but pure guess, is that it could be due to a virus of the strange class known as the "slow" viruses. One such agent was isolated some years back from the brain tissue of demented patients in a landlocked tribe in New Guinea, where most members of the group lost their minds by the age of forty to a disease known locally as "kuru." It was learned that ritual cannibalism had been a long tradition in this tribe, involving the eating of the brain of each deceased by others, including young children, and Gajdusek and his colleagues succeeded in transmitting the disease to chimpanzees by intracerebral inoculations of brain tissue. Later on, it was learned that a similar disease occurs sporadically but rarely elsewhere in the world, under the names of its discoverers, Creutzfeld and Jakob, and a similar virus has been recovered from the brain tissue of such patients.

The virus is itself the most peculiar of all forms of life. It is in the first place almost impossible to kill, resisting such traditional sterilizing procedures as boiling and exposure to alcohol, formaldehyde, and other disinfectants; indeed, the agent was recovered intact from several specimens of brain tissue that had been placed in formalin and stored in pathological museums for a decade or longer. When injected into susceptible laboratory animals (chimpanzees, guinea pigs, mice, hamsters) it produces no evidence of disease for periods of eighteen months or longer, after which the brain undergoes rapid destruction. Unlike other viruses it cannot be shown—or hasn't yet been shown—to consist of particles visible by electron microscopy, nor has anyone found nucleic acid to account for its ability to multiply many times over in the affected brain, even when the brain is known to contain a trillion infectious "units." If it is *not* made up of nucleic acid, and can nonetheless replicate itself, it will surely be the strangest of all creatures on this planet, but this is not yet known for sure; perhaps it possesses its own DNA or RNA, hidden away where it cannot be found, or it may be similar to a class of naked, very small viruses seen in plants. At present, it is a mystery.

Creutzfeld-Jakob disease is rare, accounting for only a small minority of cases of dementia in this country, but it could well be that it is not so unique as it seems. The idea that it may be related in its causative mechanism to the larger group of brain diseases labeled Alzheimer's is not outlandish. But it would still be a considerable gamble to launch a new laboratory on the hypothesis that Alzheimer's is caused by a similar slow virus, and very few such projects are planned or under way.

But quite apart from a general reluctance to take on such a long shot, there is another reason for reluctance. Even if it should turn out to be a valid notion, the work would require more time, and more money, than most biomedical scientists believe they have at their disposal these days. If you were planning to inject samples of suspected virus into an array of experimental animals, the thought of an interval of a year and a half or two years before you could reasonably expect even a preliminary, tentative answer would stay your hand, especially if you were a scientist just beginning a career. By the time you finished your first experiment you would be near the end of your first grant, most likely with nothing to show for it: no annual report, no publications, nothing to put before a faculty committee charged with considering your eligibility for promotion.

The climate is wrong for research problems like this one. Most NIH grants are awarded for periods of two to three years, most young investigators have their personal salaries paid from such grants, and there is a general, feverish sense that the research projects have to be absolutely sure things, bound to result in published papers and grant renewals by the end of the first year.

Even if an investigator decided to take a safer (and perhaps more productive) route into the problem by studying, say, the detailed characteristics of some of the already known slow viruses—"scrapie," a disease of sheep transmissible to mice, or a similar disease affecting mink brains—in hopes of learning something new about the structure or function of such viruses (if that is what they are) that might then be applied to the human disease, there is still the barrier of time and money. Each experiment requires many months of waiting, and the cost of maintaining mice or guinea pigs in good health over such long periods of time is formidable. Scientists, even those consumed by curiosity over intriguing and engrossing puzzles, tend to stay away from problems like these.

What is needed is a new kind of research support mechanism, designed specifically and selectively for the problem of senile dementia. Such a mechanism must provide money, of course, and a good deal of money, but it must also take care of the problem of time, which is peculiar to this particular biological puzzle, and it must do so in a way not available—or anyway not yet available—within the research support mechanisms of the NIH. Something of the order of ten years of guaranteed support for each laboratory is necessary, with assured increments annually indexed for inflation. At least six new laboratories should be organized and launched within the country's major research universities.

Ten years may seem an extraordinarily long period of commitment for a scientific program, and indeed it is. But it may not be long enough, and I would not argue against fifteen, or even twenty.

Now, who should foot that bill, and assume all that responsibility? Not the federal government, obviously. Governmental agencies cannot obligate themselves for such a stretch of time; they are good at thinking two years ahead, sometimes four, but no further. Who then?

I believe this is a task for the several large private foundations that have staked their mission in what is called the Health-Care Delivery System. Senile dementia is, or should be, high on their agenda of concerns, nagging away at their staffs because of the present high cost, the predictable escalation of cost in the coming decades (enough to swamp all other parts of any health-care system), and, above all, the plain fact that nothing at all can be done to alleviate this disease problem by reorganizing existing medical-care facilities or by building new ones. No training programs, no enlistment of new professionals or nonprofessionals, no rearrangements of payment mechanisms, nothing based on today's level of information about the disease can possibly be useful. The only hope lies in research, and the research will not be done—not on the scale appropriate to the problem—unless the foundations step in.

The trustees of some of the largest foundations have resolved to stay away from any involvement in basic biomedical science, in the belief that the federal government should shoulder that responsibility, and up to now they have held to that resolve, spending many millions each year on wholly commendable efforts to improve the delivery of health care. Now they have done that, and good for them. I would be glad to contribute to a banquet in their honor, to celebrate their achievements, and all my friends would come, provided that on the occasion the trustees would announce that henceforth, for the next fifteen years, one-half of their endowment income would be committed to research on dementia.

As a personal footnote, I must confess to another motive in pressing for more work on the slow virus of Creutzfeld-Jakob disease. Scientifically, it *ought* to be irresistible. The only reason it is being resisted is that too few people have had adequate long-term support to set out on such new and shifting ground. But think of the intellectual reward. To be able to catch hold of, and inspect from all sides, a living, self-replicating form of life that nobody has so far been able to see or detect by chemical methods, and one that may turn out to have its own private mechanisms for producing progeny, novel to the earth's life, should be the chance of a lifetime for any investigator. To have such a biological riddle, sitting there unsolved and neglected, is an embarrassment for biological science.

The Lie Detector

Every once in a while the reasons for discouragement about the human prospect pile up so high that it becomes difficult to see the way ahead, and it is then a great blessing to have one conspicuous and irrefutable good thing to think about ourselves, something solid enough to step onto and look beyond the pile.

Language is often useful for this, and music. A particular painting, if you have the right receptors, can lift the spirits and hold them high enough to see a whole future for the race. The sound of laughter in the distance in the dark can be a marvelous encouragement. But these are chancy stimuli, ready to work only if you happen to be ready to receive them, which takes a bit of luck.

I have been reading magazine stories about the technology of lie detection lately, and it occurs to me that this may be the thing I've been looking for, an encouragement propped up by genuine, hard scientific data. It is promising enough that I've decided to take as given what the articles say, uncritically, and to look no further. For a while, anyway.

As I understand it, a human being cannot tell a lie, even a small one, without setting off a kind of smoke alarm somewhere deep in a dark lobule of the brain, resulting in the sudden discharge of nerve impulses, or the sudden outpouring of neurohormones of some sort, or both. The outcome, recorded by the lie-detector gadgetry, is a highly reproducible cascade of changes in the electrical conductivity of the skin, the heart rate, and the manner of breathing, similar to the responses to various kinds of stress.

Lying, then, is stressful, even when we do it for protection, or relief, or escape, or profit, or just for the pure pleasure of lying and getting away with it. It is a strain, distressing enough to cause the emission of signals to and from the central nervous system warning that something has gone wrong. It is, in a pure physiological sense, an unnatural act.

Now I regard this as a piece of extraordinarily good news, meaning, unless I have it all balled up, that we are a moral species by compulsion, at least in the limited sense that we are biologically designed to be truthful to each other. Lying doesn't hurt, mind you, and perhaps you could tell lies all day and night for years on end without being damaged, but maybe not—maybe the lie detector informs us that repeated, inveterate untruthfulness will gradually undermine the peripheral vascular system, the sweat glands, the adrenals, and who knows what else. Perhaps we should be looking into the possibility of lying as an etiologic agent for some of the common human ailments still beyond explaining, recurrent head colds, for instance, or that most human of all unaccountable disorders, a sudden pain in the lower mid-back.

It makes a sort of shrewd biological sense, and might therefore represent a biological trait built into our genes, a feature of humanity as characteristic for us as feathers for birds or scales for fish, enabling us to live, at our best, the kinds of lives we are designed to live. This is, I suppose, the "sociobiological" view to take, with the obvious alternative being that we are brought up this way as children in response to the rules of our culture. But if the latter is the case, you would expect to encounter, every once in a while, societies in which the rule does not hold, and I have never heard of a culture in which lying was done by everyone as a matter of course, all life through, nor can I imagine such a group functioning successfully. Biologically speaking, there is good reason for us to restrain ourselves from lying outright to each other whenever possible. We are indeed a social species, more interdependent than the celebrated social insects; we can no more live a solitary life than can a bee; we are obliged, as a species, to rely on each other. Trust is a fundamental requirement for our kind of existence, and without it all our linkages would begin to snap loose.

The restraint is a mild one, so gentle as to be almost imperceptible. But it is there; we know about it from what we call guilt, and now we have a neat machine to record it as well.

It seems a trivial thing to have this information, but perhaps it tells us to look again, and look deeper. If we had better instruments, designed for profounder probes, we might see needles flipping, lines on charts recording quantitative degrees of meanness of spirit, or a lack of love. I do not wish for such instruments, I hope they will never be constructed; they would somehow belittle the issues involved. It is enough, quite enough, to know that we cannot even tell a plain untruth, betray a trust, without scaring some part of our own brains. I'd rather guess at the rest.

There is, of course, one problem that will have to be straightened out sooner or later by medicine, duty-bound. It concerns placebos. The sugar pill is sometimes indispensable in therapy, powerfully reassuring, but it is essentially a little white lie. If you wired up the average good internist in the act of writing a prescription, would the needles go flying?

Let others go to work on the scientific side issues, of which there are probably many. Is there a skin secretion, a pheromone, secreted in the process? Can a trained tracking hound smell the altered skin of a liar? Is the total absence of this secretion the odor of sanctity? I can think of any number of satisfying experiments that someone ought to be doing, but I confess to a serious misgiving about the possible misuses of the sort of knowledge I have in mind. Supposing it were found that there is indeed a special pentapeptide released into the blood on the telling of a lie, or some queer glycolipid in the sweat of one's palms, or, worst of all, something chemically detectable in balloons of exhaled breath. The next thing to happen would surely be new industries in Texas and Japan for the manufacture of electronic sensing devices to be carried in one's pocket, or perhaps worn conspicuously on one's sleeve depending on the consumer's particular need. Governments would become involved, sooner or later, and the lawyers and ethicists would have one field day after another. Before long we would stop speaking to each other, television would be abolished as a habitual felon, politicians would be confined by house arrest, and civilization would come to a standstill.

Come to think of it, you might not have to do any of the research on human beings after all, which I find a relaxing thought. Animals, even plants, lie to each other all the time, and we could restrict the research to them, putting off the real truth about ourselves for the several centuries we need to catch our breath. What is it that enables certain flowers to resemble nubile insects, or opossums to play dead, or female fireflies to change the code of their flashes in order to attract, and then eat, males of a different species? What about those animals that make their livings by deception—the biological mimics, the pretenders, the fish dangling bits of their flesh as bait in front of their jaws, the malingering birds limping along to lie about the location of their nests, the peacock, who is surely not conceivably all that he claims to be? It is a rich field indeed, open to generations of graduate students in the years ahead, risk-free. All we need is to keep telling ourselves that this is not a human problem, to understand that we have evolved beyond mendacity except under extraordinary conditions, and to stay clear of the instruments.

It would be safe enough for the scientists themselves, of course, because good science depends on truth-telling, and we should be willing to wear detectors on the lapels of our white coats all day long. I have only one small reservation about this. Scientists do have a tendency to vanity—some of the best ones are vanity-prone—and there is probably a mechanism at work here with a fundamental connection to lying. Perhaps this is one kind of human experimentation that ought to be done early on, if it can pass review by the local ethics review board: catch hold of an eminent researcher at the moment when he is involved in a press conference, looking and sounding for all the world like the greatest thing since the invention of the nucleated cell, and hook him up to the machine, or stick a sensor on his necktie. Then we could learn how to control the work for background noise, and move on to the insects.

I don't want to go over this again. I didn't write any of the above.

Some Scientific Advice

It was good news when a Science Adviser was appointed and installed in the White House, put there with the explicit understanding that his job is not to represent the scientific community as an advocate but to provide the President with the best-informed and most objective counsel available for the formulation of a national science policy. We have had Science Advisers in the White House before, but never a recognizable science policy. It is an ambitious undertaking, and the Adviser will need all the advice he can get. No single scientist, or any full-time staff assigned to his office, can possibly appraise and evaluate the progress and problems all along the immense frontier of today's science in this country, let alone in the rest of the world. He will need the services of expert committees and panels representing the various broad fields of science, industry, and education. He will need as well good advice from thoughtful, sagacious citizens who have no connection at all with science or scientists. Something like the structure of the President's Scientific Advisory Committee (PSAC), in useful existence until evicted in the late Nixon years, will be indispensable for the Science Adviser's work, and it is probable that sooner or later such a body will be created politically neutral, one hopes, made up of people who can agree not to act as special pleaders for the constituencies of science and technology, but to provide unbiased advice for the country's research effort in the years ahead.

Having served as a member of PSAC for four years in the late 1960s and early 1970s, I am aware of the difficulties involved in objectivity. The biologists will want more biomedical science, the physicists will want more physics and more big instruments, the social scientists will hope at least to ensure the survival of their disciplines in hard times, the industrialists will demand more applied research, and the citizens-at-large will want to make sure that the health, safety, and well-being of ordinary people are enhanced by science and not, as some apprehend, placed more in jeopardy. Somewhere offstage, the military and intelligence communities will be wanting things to go their way in research.

But it begins already to look like an entirely new set of puzzles for which the administration will be seeking advice. In the old days, the early 1970s, the main task for the Science Adviser and his advisers was to identify the most important national problems for which a better kind of scientific study might be useful, and then to press for expansion of that sort of research, whatever. There were opportunities all over the place, and the PSAC panels fired off one report after another, always recommending more research. Now things are different. Nobody in the upper reaches of government is likely to be looking around for new ways to spend money. It is much more probable that the Science Adviser will look over the country's scientific endeavor with an eye out for expenditures that can be reduced or eliminated.

The word is out that the United States cannot do all the science that needs doing. Instead of trying to explore all aspects of biology or physics or chemistry or behavioral science, we are told that a more limited agenda must be arranged. The problem will be to identify a finite number of surefire areas of science, and concentrate our efforts on these in hopes of achieving prompt and profitable payoffs. Energy is an obvious candidate, agriculture another, biological engineering another.

If this is the way things are to go, the Adviser and his committees are likely to have a dreary time of it. Trying to make guesses at the future in research is an easy enough job if you are talking about matters of some certainty—the likelihood of getting certain proteins more cheaply from bacteria by the recombinant-DNA technique, for example—but trying to guess which lines of fundamental science are *unlikely* to yield profitable knowledge is quite another matter. I cannot imagine a more depressing undertaking for a committee, no matter how bright and stimulating its members.

What branches of science should we now give up on, or turn over to the rising generations of increasingly adept investigators in Europe or the United Kingdom or Japan or the Soviet Union? Or, within those branches, which particular lines of investigation should we set aside on grounds that their prospects for a short-term payoff are too marginal for an investment? I can see ways of answering questions like these in applied research: obviously solar energy versus nuclear fusion versus conservation versus fossil fuels versus hydrogen are items that can be argued over in terms of the dollars and years needed for research, and reasonably intelligent appraisals can be arrived at.

But what can be said about the future yields from *basic* research, in any field of science? No Science Adviser, nor any committee, has ever succeeded in forecasting the future outcome of this kind of scientific endeavor. It is in the nature of basic research that the future is unknowable until it happens. No committee could have sat around a table in the early 1950s and predicted that the pursuit of work in solid-state physics would produce the microchip. All the world's molecular biologists of the 1960s, assembled in any conference hall, could not have imagined that the strings of genes of one species could soon be inserted into the DNA of a totally different species for manufacturing salable products. Nor can any group of our best cell biologists, hot on the trail of mechanisms responsible for cancer, predict with any confidence which particular line of research carries a higher probability of success than another line.

What are we talking about, anyway? Is this an argument over costs, and is the country so near to being broke that the President must be advised to reduce the national effort in science? I cannot imagine it. This is a special intellectual knack, a sort of national, natural gift, in which the United States excels. It is one of the things that Americans and their institutions—mainly their talented universities—are really good at. And, if I may say so at a time when every federal penny is to be watched and pinched, good basic science comes relatively cheap. As a percentage of the gross national product, the amount being spent now on basic science is so small that it would go undiscovered if incorporated into the budget of the Department of Defense. We can afford to spend more than we spend today.

My unsolicited and perhaps unwelcome advice to the Adviser would be to plunge, to splurge, to cut back nowhere, to encourage the doing of basic research wherever the questions seem engrossing and fascinating, and not to *think* of excluding any field, never mind the cost. It is the best investment, short-term or long-term, that the country can make.

The Attic of the Brain

My parents' house had an attic, the darkest and strangest part of the building, reachable only by placing a stepladder beneath the trapdoor and filled with unidentifiable articles too important to be thrown out with the trash but no longer suitable to have at hand. This mysterious space was the memory of the place. After many years all the things deposited in it became, one by one, lost to consciousness. But they were still there, we knew, safely and comfortably stored in the tissues of the house.

These days most of us live in smaller, more modern houses or in apartments, and attics have vanished. Even the deep closets in which we used to pile things up for temporary forgetting are rarely designed into new homes.

Everything now is out in the open, openly acknowledged and displayed, and whenever we grow tired of a memory, an old chair, a trunkful of old letters, they are carted off to the dump for burning.

This has seemed a healthier way to live, except maybe for the smoke—everything out to be looked at, nothing strange hidden under the roof, nothing forgotten because of no place left in impenetrable darkness to forget. Openness is the new life-style, no undisclosed belongings, no private secrets. Candor is the rule in architecture. The house is a machine for living, and what kind of a machine would hide away its worn-out, obsolescent parts?

But it is in our nature as human beings to clutter, and we hanker for places set aside, reserved for storage. We tend to accumulate and outgrow possessions at the same time, and it is an endlessly discomforting mental task to keep sorting out the ones to get rid of. We might, we think, remember them later and find a use for them, and if they are gone for good, off to the dump, this is a source of nervousness. I think it may be one of the reasons we drum our fingers so much these days.

We might take a lesson here from what has been learned about our brains in this century. We thought we discovered, first off, the attic, although its existence has been mentioned from time to time by all the people we used to call great writers. What we really found was the trapdoor and a stepladder, and off we clambered, shining flashlights into the corners, vacuuming the dust out of bureau drawers, puzzling over the names of objects, tossing them down to the floor below, and finally paying around fifty dollars an hour to have them carted off for burning.

After several generations of this new way of doing things

we took up openness and candor with the febrile intensity of a new religion, everything laid out in full view, and as in the design of our new houses it seemed a healthier way to live, except maybe again for smoke.

And now, I think, we have a new kind of worry. There is no place for functionless, untidy, inexplicable notions, no dark comfortable parts of the mind to hide away the things we'd like to keep but at the same time forget. The attic is still there, but with the trapdoor always open and the stepladder in place we are always in and out of it, flashing lights around, naming everything, unmystified.

I have an earnest proposal for psychiatry, a novel set of therapeutic rules, although I know it means waiting in line.

Bring back the old attic. Give new instructions to the patients who are made nervous by our times, including me, to make a conscious effort to hide a reasonable proportion of thought. It would have to be a gradual process, considering how far we have come in the other direction talking, talking all the way. Perhaps only one or two thoughts should be repressed each day, at the outset. The easiest, gentlest way might be to start with dreams, first by forbidding the patient to mention any dream, much less to recount its details, then encouraging the outright forgetting that there was a dream at all, remembering nothing beyond the vague sense that during sleep there had been the familiar sound of something shifting and sliding, up under the roof.

We might, in this way, regain the kind of spontaneity and zest for ideas, things popping into the mind, uncontrollable and ungovernable thoughts, the feel that this notion is somehow connected unaccountably with that one.

We could come again into possession of real memory, the kind of memory that can come only from jumbled forgotten furniture, old photographs, fragments of music.

It has been one of the great errors of our time to think that by thinking about thinking, and then talking about it, we could possibly straighten out and tidy up our minds. There is no delusion more damaging than to get the idea in your head that you understand the functioning of your own brain. Once you acquire such a notion, you run the danger of moving in to take charge, guiding your thoughts, shepherding your mind from place to place, *controlling* it, making lists of regulations. The human mind is not meant to be governed, certainly not by any book of rules yet written; it is supposed to run itself, and we are obliged to follow it along, trying to keep up with it as best we can. It is all very well to be aware of your awareness, even proud of it, but never try to operate it. You are not up to the job.

I leave it to the analysts to work out the techniques for doing what now needs doing. They are presumably the professionals most familiar with the route, and all they have to do is turn back and go the other way, session by session, step by step. It takes a certain amount of hard swallowing and a lot of revised jargon, and I have great sympathy for their plight, but it is time to reverse course.

If after all, as seems to be true, we are endowed with unconscious minds in our brains, these should be regarded as normal structures, installed wherever they are for a purpose. I am not sure what they are built to contain, but as a biologist, impressed by the usefulness of everything alive, I would take it for granted that they are useful, probably indispensable organs of thought. It cannot be a bad thing to own one, but I would no more think of meddling with it than trying to exorcise my liver, an equally mysterious apparatus. Until we know a lot more, it would be wise, as we have learned from other fields in medicine, to let them be, above all not to interfere. Maybe, even—and this is the notion I wish to suggest to my psychiatric friends—to stock them up, put more things into them, make *use* of them. Forget whatever you feel like forgetting. From time to time, practice *not* being open, discover new things *not* to talk about, learn reserve, hold the tongue. But above all, develop the human talent for forgetting words, phrases, whole unwelcome sentences, all experiences involving wincing. If we should ever lose the loss of memory, we might lose as well that most attractive of signals ever flashed from the human face, the blush. If we should give away the capacity for embarrassment, the touch of fingertips might be the next to go, and then the suddenness of laughter, the unaccountable sure sense of something gone wrong, and, finally, the marvelous conviction that being human is the best thing to be.

Attempting to operate one's own mind, powered by such a magical instrument as the human brain, strikes me as rather like using the world's biggest computer to add columns of figures, or towing a Rolls-Royce with a nylon rope.

I have tried to think of a name for the new professional activity, but each time I think of a good one I forget it before I can get it written down. Psychorepression is the only one I've hung on to, but I can't guess at the fee schedule.

Humanities and Science

Lord Kelvin was one of the great British physicists of the late nineteenth century, an extraordinarily influential figure in his time, and in some ways a paradigm of conventional, established scientific leadership. He did a lot of good and useful things, but once or twice he, like Homer, nodded. The instances are worth recalling today, for we have nodders among our scientific eminences still, from time to time, needing to have their elbows shaken.

On one occasion, Kelvin made a speech on the overarching importance of numbers. He maintained that no observation of nature was worth paying serious attention to unless it could be stated in precisely quantitative terms. The numbers were the final and only test, not only of truth but about meaning as well. He said, "When you can measure what you are speaking about, and express it in numbers, you know something about it. But when you cannot—your knowledge is of a meagre and unsatisfactory kind."

But, as at least one subsequent event showed, Kelvin may have had things exactly the wrong way round. The task of converting observations into numbers is the hardest of all, the last task rather than the first thing to be done, and it can be done only when you have learned, beforehand, a great deal about the observations themselves. You can, to be sure, achieve a very deep understanding of nature by quantitative measurement, but you must know what you are talking about before you can begin applying the numbers for making predictions. In Kelvin's case, the problem at hand was the age of the earth and solar system. Using what was then known about the sources of energy and the loss of energy from the physics of that day, he calculated that neither the earth nor the sun were older than several hundred million years. This caused a considerable stir in biological and geological circles, especially among the evolutionists. Darwin himself was distressed by the numbers; the time was much too short for the theory of evolution. Kelvin's figures were described by Darwin as one of his "sorest troubles."

T. H. Huxley had long been aware of the risks involved in premature extrapolations from mathematical treatment of biological problems. He said, in an 1869 speech to the Geological Society concerning numbers, "This seems to be one of the many cases in which the admitted accuracy of mathematical processes is allowed to throw a wholly inadmissible appearance of authority over the results obtained by them. . . . As the grandest mill in the world will not extract wheat flour from peascods, so pages of formulas will not get a definite result out of loose data."

The trouble was that the world of physics had not moved fast enough to allow for Kelvin's assumptions. Nuclear fusion and fission had not yet been dreamed of, and the true age of the earth could not even be guessed from the data in hand. It was not yet the time for mathematics in this subject.

There have been other examples, since those days, of the folly of using numbers and calculations uncritically. Kelvin's own strong conviction that science could not be genuine science without measuring things was catching. People in other fields of endeavor, hankering to turn their disciplines into exact sciences, beset by what has since been called "physics envy," set about converting whatever they knew into numbers and thence into equations with predictive pretensions. We have it with us still, in economics, sociology, psychology, history, even, I fear, in English-literature criticism and linguistics, and it frequently works, when it works at all, with indifferent success. The risks of untoward social consequences in work of this kind are considerable. It is as important—and as hard—to learn *when* to use mathematics as *how* to use it, and this matter should remain high on the agenda of consideration for education in the social and behavioral sciences.

Of course, Kelvin's difficulty with the age of the earth was an exceptional, almost isolated instance of failure in quantitative measurement in nineteenth-century physics. The instruments devised for aproaching nature by way of physics became increasingly precise and powerful, carrying the field through electromagnetic theory, triumph after triumph, and setting the stage for the great revolution of twentieth-century physics. There is no doubt about it: measurement works when the instruments work, and when you have a fairly clear idea of what it is that is being measured, and when you know what to do with the numbers when they tumble out. The system for gaining information and comprehension about nature works so well, indeed, that it carries another hazard: the risk of convincing yourself that you know everything.

Kelvin himself fell into this trap toward the end of the century. (I don't mean to keep picking on Kelvin, who was a very great scientist; it is just that he happened to say a couple of things I find useful for this discussion.) He stated, in a summary of the achievements of nineteenth-century physics, that it was an almost completed science; virtually everything that needed knowing about the material universe had been learned; there were still a few anomalies and inconsistencies in electromagnetic theory, a few loose ends to be tidied up, but this would be done within the next several years. Physics, in these terms, was not a field any longer likely to attract, as it previously had, the brightest and most imaginative young brains. The most interesting part of the work had already been done. Then, within the next decade, came radiation, Planck, the quantum, Einstein, Rutherford, Bohr, and all the rest—quantum mechanics—and the whole field turned over and became a brand-new sort of human endeavor, still now, in the view of many physicists, almost a full century later, a field only at its beginnings.

But even today, despite the amazements that are turning up in physics each year, despite the jumps taken from the smallest parts of nature—particle physics—to the largest of all—the cosmos itself—the impression of science that the public gains is rather like the impression left in the nineteenth-century public mind by Kelvin. Science, in this view, is first of all a matter of simply getting all the numbers together. The numbers are sitting out there in nature, waiting to be found, sorted and totted up. If only they had enough robots and enough computers, the scientists could go off to the beach and wait for their papers to be written for them. Second of all, what we know about nature today is pretty much the whole story: we are very nearly home and dry. From here on, it is largely a problem of tying up loose ends, tidying nature up, getting the files in order. The only real surprises for the future—and it is about those that the public is becoming more concerned and apprehensive—are the technological applications that the scientists may be cooking up from today's knowledge.

I suggest that the scientific community is to blame. If there are disagreements between the world of the humanities and the scientific enterprise as to the place and importance of science in a liberal-arts education, and the role of science in twentieth-century culture, I believe that the scientists are themselves responsible for a general misunderstanding of what they are really up to.

Over the past half century, we have been teaching the sciences as though they were the same academic collection of cut-and-dried subjects as always, and—here is what has really gone wrong—as though they would always be the same. The teaching of today's biology, for example, is pretty much the same kind of exercise as the teaching of Latin was when I was in high school long ago. First of all, the fundamentals, the underlying laws, the essential grammar, and then the reading of texts. Once mastered, that is that: Latin is Latin and forever after will be Latin. And biology is precisely biology, a vast array of hard facts to be learned as fundamentals, followed by a reading of the texts.

Moreover, we have been teaching science as though its facts were somehow superior to the facts in all other scholarly disciplines, more fundamental, more solid, less subject to subjectivism, immutable. English literature is not just one way of thinking, it is all sorts of ways. Poetry is a moving target. The facts that underlie art, architecture, and music are not really hard facts, and you can change them any way you like by arguing about them, but science is treated as an altogether different kind of learning: an unambiguous, unalterable, and endlessly useful display of data needing only to be packaged and installed somewhere in one's temporal lobe in order to achieve a full understanding of the natural world.

And it is, of course, not like this at all. In real life, every field of science that I can think of is incomplete, and most of them—whatever the record of accomplishment over the past two hundred years—are still in the earliest stage of their starting point. In the fields I know best, among the life sciences, it is required that the most expert and sophisticated minds be capable of changing those minds, often with a great lurch, every few years. In some branches of biology the mind-changing is occurring with accelerating velocities. The next week's issue of any scientific journal can turn a whole field upside down, shaking out any number of immutable ideas and installing new bodies of dogma, and this is happening all the time. It is an almost everyday event in physics, in chemistry, in materials research, in neurobiology, in genetics, in immunology. The hard facts tend to soften overnight, melt away, and vanish under the pressure of new hard facts, and the interpretations of what appear to be the most solid aspects of nature are subject to change, now more than at any other time in history. The conclusions reached in science are always, when looked at closely, far more provisional and tentative than are most of the assumptions arrived at by our colleagues in the humanities.

The running battle now in progress between the sociobiologists and the antisociobiologists is a marvel for students to behold, close up. To observe, in open-mouthed astonishment, the polarized extremes, one group of highly intelligent, beautifully trained, knowledgeable, and imaginative scientists maintaining that all sorts of behavior, animal and human, are governed exclusively by genes, and another group of equally talented scientists saying precisely the opposite and asserting that all behavior is set and determined by the environment, or by culture, and both sides brawling in the pages of periodicals such as *The New York Review of Books*, is an educational experience that no college student should be allowed to miss. The essential lesson to be learned has nothing to do with the relative validity of the facts underlying the argument, it is the argument itself that is the education: we do not yet know enough to settle such questions.

It is true that at any given moment there is the appearance of satisfaction, even self-satisfaction, within every scientific discipline. On any Tuesday morning, if asked, a good working scientist will gladly tell you that the affairs of the field are nicely in order, that things are finally looking clear and making sense, and all is well. But come back again, on another Tuesday, and he may let you know that the roof has just fallen in on his life's work, that all the old ideas—last week's ideas in some cases—are no longer good ideas, that something strange has happened.

It is the very strangeness of nature that makes science engrossing. That ought to be at the center of science teaching. There are more than seven-times-seven types of ambiguity in science, awaiting analysis. The poetry of Wallace Stevens is crystal-clear alongside the genetic code.

I prefer to turn things around in order to make precisely the opposite case. Science, especially twentieth-century science, has provided us with a glimpse of something we never really knew before, the revelation of human ignorance. We have been used to the belief, down one century after another, that we more or less comprehend everything bar one or two mysteries like the mental processes of our gods. Every age, not just the eighteenth century, regarded itself as the Age of Reason, and we have never lacked for explanations of the world and its ways. Now, we are being brought up short, and this has been the work of science. We have a wilderness of mystery to make our way through in the centuries ahead, and we will need science for this but not science alone. Science will, in its own time, produce the data and some of the meaning in the data, but never the full meaning. For getting a full grasp, for perceiving real significance when significance is at hand, we shall need minds at work from all sorts of brains outside the fields of science, most of all the brains of poets, of course, but also those of artists, musicians, philosophers, historians, writers in general.

It is primarily because of this need that I would press for changes in the way science is taught. There is a need to teach the young people who will be doing the science themselves, but this will always be a small minority among us. There is a deeper need to teach science to those who will be needed for thinking about it, and this means pretty nearly everyone else, in hopes that a few of these people—a much smaller minority than the scientific community and probably a lot harder to find—will, in the thinking, be able to imagine new levels of meaning that are likely to be lost on the rest of us.

In addition, it is time to develop a new group of professional thinkers, perhaps a somewhat larger group than the working scientists, who can create a discipline of scientific criticism. We have had good luck so far in the emergence of a few people ranking as philosophers of science and historians and journalists of science, and I hope more of these will be coming along, but we have not yet seen a Ruskin or a Leavis or an Edmund Wilson. Science needs critics of this sort, but the public at large needs them more urgently.

I suggest that the introductory courses in science, at all levels from grade school through college, be radically revised. Leave the fundamentals, the so-called basics, aside for a while, and concentrate the attention of all students on the things that are *not* known. You cannot possibly teach quantum mechanics without mathematics, to be sure, but you can describe the strangeness of the world opened up by quantum theory. Let it be known, early on, that there are deep mysteries, and profound paradoxes, revealed in their distant outlines, by the quantum. Let it be known that these can be approached more closely, and puzzled over, once the language of mathematics has been sufficiently mastered.

Teach at the outset, before any of the fundamentals, the still imponderable puzzles of cosmology. Let it be known, as clearly as possible, by the youngest minds, that there are some things going on in the universe that lie beyond comprehension, and make it plain how little is known.

Do not teach that biology is a useful and perhaps profitable science; that can come later. Teach instead that there are structures squirming inside all our cells, providing all the energy for living, that are essentially foreign creatures, brought in for symbiotic living a billion or so years ago, the lineal descendants of bacteria. Teach that we do not have the ghost of an idea how they got there, where they came from, or how they evolved to their present structure and function. The details of oxidative phosphorylation and photosynthesis can come later.

Teach ecology early on. Let it be understood that the earth's life is a system of interliving, interdependent creatures, and that we do not understand at all how it works. The earth's environment, from the range of atmospheric gases to the chemical constituents of the sea, has been held in an almost unbelievably improbable state of regulated balance since life began, and the regulation of stability and balance is accomplished solely by the life itself, like the internal environment of an immense organism, and we do not know how *that* one works, even less what it means. Teach that.

Go easy, I suggest, on the promises sometimes freely offered by science. Technology relies and depends on science these days, more than ever before, but technology is nothing like the first justification for doing research, nor is it necessarily an essential product to be expected from science. Public decisions about what to have in the way of technology are totally different problems from decisions about science, and the two enterprises should not be tangled together. The central task of science is to arrive, stage by stage, at a clearer comprehension of nature, but this does not mean, as it is sometimes claimed to mean, a search for mastery over nature. Science may provide us, one day, with a better understanding of ourselves, but never, I hope, with a set of technologies for doing something or other to improve ourselves. I am made nervous by assertions that human consciousness will someday be unraveled by research, laid out for close scrutiny like the workings of a computer, and then, *and then*! I hope with some fervor that we can learn a lot more than we now know about the human mind, and I see no reason why this strange puzzle should remain forever and entirely beyond us. But I would be deeply disturbed by any prospect that we might use the new knowledge in order to begin doing something about it, to improve it, say. This is a different matter from searching for information to use against schizophrenia or dementia, where we are badly in need of technologies, indeed likely one day to be sunk without them. But the ordinary, everyday, more or less normal human mind is too marvelous an instrument ever to be tampered with by anyone, science or no science.

The education of humanists cannot be regarded as complete, or even adequate, without exposure in some depth to where things stand in the various branches of science, and particularly, as I have said, in the areas of our ignorance.

This does not mean that I know how to go about doing it, nor am I unaware of the difficulties involved. Physics professors, most of them, look with revulsion on assignments to teach their subject to poets. Biologists, caught up by the enchantment of their new power, armed with flawless instruments to tell the nucleotide sequences of the entire human genome, nearly matching the physicists in the precision of their measurements of living processes, will resist the prospect of broad survey courses; each biology professor will demand that any student in his path must master every fine detail within that professor's research program. The liberal-arts faculties, for their part, will continue to view the scientists with suspicion and apprehension. "What do the scientists want?" asked a Cambridge professor in Francis Cornford's wonderful *Microcosmographia Academica.* "Everything that's going," was the quick answer. That was back in 1912, and universities haven't much changed.

The worst thing that has happened to science education is that the great fun has gone out of it. A very large number of good students look at it as slogging work to be got through on the way to medical school. Others look closely at the premedical students themselves, embattled and bleeding for grades and class standing, and are turned off. Very few see science as the high adventure it really is, the wildest of all explorations ever undertaken by human beings, the chance to catch close views of things never seen before, the shrewdest maneuver for discovering how the world works. Instead, they become baffled early on, and they are misled into thinking that bafflement is simply the result of not having learned all the facts. They are not told, as they should be told, that everyone else—from the professor in his endowed chair down to the platoons of postdoctoral students in the laboratory all night—is baffled as well. Every important scientific advance that has come in looking like an answer has turned, sooner or later—usually sooner—into a question. And the game is just beginning.

An appreciation of what is happening in science today, and of how great a distance lies ahead for exploring, ought to be one of the rewards of a liberal-arts education. It ought to be a good in itself, not something to be acquired on the way to a professional career but part of the cast of thought needed for getting into the kind of century that is now just down the road. Part of the intellectual equipment of an educated person, however his or her time is to be spent, ought to be a feel for the queernesses of nature, the inexplicable things.

And maybe, just maybe, a new set of courses dealing systematically with ignorance in science might take hold. The scientists might discover in it a new and subversive technique for catching the attention of students driven by curiosity, delighted and surprised to learn that science is exactly as Bush described it: an "endless frontier." The humanists, for their part, might take considerable satisfaction watching their scientific colleagues confess openly to not knowing everything about everything. And the poets, on whose shoulders the future rests, might, late nights, thinking things over, begin to see some meanings that elude the rest of us. It is worth a try.

On Matters of Doubt

The "two-cultures" controversy of several decades back has quieted down some, but it is still with us, still unsettled because of the polarized views set out by C. P. Snow at one polemical extreme and by F. R. Leavis at the other; these remain as the two sides of the argument. At one edge, the humanists are set up as knowing, and wanting to know, very little about science and even less about the human meaning of contemporary science; they are, so it goes, antiscientific in their prejudice. On the other side, the scientists are served up as a bright but illiterate lot, well-read in nothing except science, even, as Leavis said of Snow, incapable of writing good novels. The humanities are presented in the dispute as though made up of imagined unverifiable notions about human behavior, unsubstantiated stories cooked up by poets and novelists, while the sciences deal parsimoniously with lean facts, hard data, incontrovertible theories, truths established beyond doubt, the unambiguous facts of life.

The argument is shot through with bogus assertions and false images, and I have no intention of becoming entrapped in it here, on one side or the other. Instead, I intend to take a stand in the middle of what seems to me a muddle, hoping to confuse the argument by showing that there isn't really any argument in the first place. To do this, I must try to show that there is in fact a solid middle ground to stand on, a shared common earth beneath the feet of all the humanists and all the scientists, a single underlying view of the world that drives all scholars, whatever their discipline—whether history or structuralist criticism or linguistics or quantum chromodynamics or astrophysics or molecular genetics.

There is, I think, such a shared view of the world. It is called *bewilderment.* Everyone knows this, but it is not much talked about; bewilderment is kept hidden in the darkest closets of all our institutions of higher learning, repressed whenever it seems to be emerging into public view, sometimes glimpsed staring from attic windows like a mad cousin of learning. It is the family secret of twentieth-century science, and of twentieth-century arts and letters as well. Human knowledge doesn't stay put. What we have been learning in our time is that we really do not understand this place or how it works, and we comprehend our own selves least of all. And the more we learn, the more we are—or ought to be—dumbfounded.

It is the greatest fun to be bewildered, but only when there lies ahead the sure certainty of having things straightened out, and soon. It is like a marvelous game, provided you have some way of keeping score, and this is what seems to be lacking in our time. It is confusing, and too many of us are choosing not to play, settling back with whatever straws of fixed knowledge we can lay hands on, denying bewilderment, pretending one conviction or another, nodding our heads briskly at whatever we prefer to believe, staying away from the ambiguity of being.

We would be better off if we had never invented the terms "science" and "humanities" and then set them up as if they represented two different kinds of intellectual enterprise. I cannot see why we ever did this, but we did. Now, to make matters worse, we have these two encampments not only at odds but trying to swipe problems from each other. The historians, some of them anyway, want to be known as social scientists and solve the ambiguities of history by installing computers in all their offices; the deconstructionists want to become the ultimate scientists of poetry, looking at every word in a line with essentially the reductionist attitude of particle physicists in the presence of atoms, but still unaware of the uncertainty principle that governs any good poem: not only can the observer change the thing observed, he can even destroy it. The biologists have invaded all aspects of human behavior with equations to explain away altruism and usefulness by totting up the needs of genes; the sociobiologists are becoming humanists manqué, swept off their feet by ants. The physicists, needing new terms for their astonishments, borrow "quarks" from Joyce and label precisely quantitative aspects of matter almost dismissively with poetically allusive words like "strangeness," "color," and "flavor"; soon some parts of the universe will begin to "itch."

We have, to be sure, learned enough to know better than to say some things, about letters and about science, but we are still too reticent about our ignorance. Most things in the world are unsettling and bewildering, and it is a mistake to try to explain them away; they are there for marveling at and wondering at, and we should be doing more of this.

I do not mean to suggest that we are surrounded by unknowable things. Indeed, I cannot imagine any sorts of questions to be asked about ourselves or about nature that cannot sooner or later be answered, given enough time. I do admit to worrying, late at night, about that matter of time: obviously we will have to get rid of modern warfare and quickly, or else we will end up, with luck, throwing spears and stones at each other. We could, without luck, run out of time in what is left of this century and then, by mistake, finish the whole game off by upheaving the table, ending life for everything except the bacteria, maybe—with enough radiation, even them. If you are given to fretting about what is going on in the minds of the young people in our schools, or on the streets of Zurich or Paris or Sydney or Tokyo or wherever, give a thought to the idea of impermanence for a whole species—*ours*—and the risk of earthly incandescence; it is a brand-new idea, never before confronted as a reality by any rising generation of human beings.

I have an idea, as an aside. Why not agree with the Russians about just one technological uniformity to be installed, at small cost, in all the missiles, theirs and ours: two small but comfortable chambers added to every vehicle before firing, one for a prominent diplomat selected by the other side, one for a lawyer selected at random? It might be a beginning.

Here's a list of things, taken more or less at random, that we do not understand:

I am entitled to say, if I like, that awareness exists in all the individual creatures on the planet—worms, sea urchins, gnats, whales, subhuman primates, superprimate humans, the lot. I can say this because we do not know what we are talking about; consciousness is so much a total mystery for our own species that we cannot begin to guess about its existence in others. I can say that bird song is the music made by songbirds for their own pleasure, pure fun, also for ours, and it is only a piece of good fortune that the music turns out to be handy for finding mates for breeding or setting territorial markers. I can say, if I like, that social insects behave like the working parts of an immense central nervous system: the termite colony is an enormous brain on millions of legs; the individual termite is a mobile neurone. This would mean that there is such a phenomenon as collective thinking, which goes on whenever sufficient numbers of creatures are sufficiently connected to one another, and it would also mean that we humans could do the same trick if we tried, and perhaps we've already done it, over and over again, in the making of language and the meditative making (for which the old Greek word *poesis* is best) of metaphors. I can even assert out loud that we are, as a species, held together by something like affection (what the physicists might be calling a "weak force") and by something like love (a "strong force"), and nobody can prove that I'm wrong. I can dismiss all the evidence piling up against such an idea, all our destructiveness and cantankerousness, as error, error-proneness, built into our species to allow more flexibility of choice, and nobody can argue me out of this unless I choose to wander off to another point of view.

I am inclined to assert, unconditionally, that there is one central, universal aspect of human behavior, genetically set by our very nature, biologically governed, driving each of us along. Depending on how one looks at it, it can be defined as the urge to be useful. This urge drives society along, sets our behavior as individuals and in groups, invents all our myths, writes our poetry, composes our music.

It is not easy to be a social species and, at the same time, such a juvenile, almost brand-new species, milling around in groups, trying to construct a civilization that will last. Being useful is easy for an ant: you just wait for the right chemical signal, at the right stage of the construction of the hill, and then you go looking for a twig of exactly the right size for that stage and carry it back, up the flank of the hill, and put it in place, and then you go and do that thing again. An ant can dine out on his usefulness, all his life, and never get it wrong.

It is a different problem for us, carrying such risks of doing it wrong, getting the wrong twig, losing the hill, not even recognizing, yet, the outline of the hill. We are beset by strings of DNA, immense arrays of genes, instructing each of us to be helpful, impelling us to try our whole lives to be useful, but never telling us how. The instructions are not coded out in anything like an operator's manual; we have to make guesses all the time. The difficulty is increased when groups of us are set to work together; I have seen, and sat on, numberless committees, not one of which intended anything other than great merit, feckless all. Larger collections of us—cities, for instance—hardly ever get anything right. And, of course, there is the modern nation, probably the most stupefying example of biological error since the age of the great reptiles, wrong at every turn, but always felicitating itself loudly on its great value. It is a biological problem, as much so as a coral reef or a rain forest, but such things as happen to human nations could never happen in a school of fish. It is, when you think about it, a humiliation, but then "humble" and "human" are cognate words. We are smarter than the fish, but their instructions come along in their eggs; ours we are obliged to figure out, and we are, in this respect, slow learners.

The sciences and the humanities are all of a piece, one and the same kind of work for the human brain, done by launching guesses and finding evidence to back up the guesses. The methods and standards are somewhat different, to be sure. It is easier to prove that something is so in science than it is to make an assertion about Homer or Cézanne or Wallace Stevens and have it stand up to criticism from all sides, harder still to *be* Homer or Cézanne or Stevens, but the game is the same game. The hardest task for the scientists, hardly yet begun, is to find out what their findings may mean, deep inside, and how one piece of solid information, firmly established by experimentation and confirmation, fits with that unlike piece over there. The natural world is all of a piece, we all know this in our bones, but we have a long, long way to go before we will see how the connections are made.

If you are looking about for really profound mysteries, essential aspects of our existence for which neither the sciences nor the humanities can provide any sort of explanation, I suggest starting with music. The professional musicologists, tremendous scholars all, for whom I have the greatest respect, haven't the ghost of an idea about what music is, or why we make it and cannot be human without it, or even—and this is the telling point—how the human mind makes music on its own, before it is written down and played. The biologists are no help here, nor the psychologists, nor the physicists, nor the philosophers, wherever they are these days. Nobody can explain it. It is a mystery, and thank goodness for that. The

Brandenburgs and the late quartets are not there to give us assurances that we have arrived; they carry the news that there are deep centers in our minds that we know nothing about except that they are there.

The thing to do, to get us through the short run, the years just ahead, is to celebrate our ignorance. Instead of presenting the body of human knowledge as a mountainous structure of coherent information capable of explaining everything about everything if we could only master all the details, we should be acknowledging that it is, in real life, still a very modest mound of puzzlements that do not fit together at all. As a species, the thing we are biologically good at is learning new things, thanks to our individual large brains and thanks above all to the gift of speech that connects them, one to another. We can take some gratification at having come a certain distance in just the few thousand years of our existence as language users, but it should be a deeper satisfaction, even an exhilaration, to recognize that we have such a distance still to go. Get us through the next few years, I say, just get us safely out of this century and into the next, and then watch what we can do.

Late Night Thoughts on Listening to Mahler's Ninth Symphony

I cannot listen to Mahler's Ninth Symphony with anything like the old melancholy mixed with the high pleasure I used to take from this music. There was a time, not long ago, when what I heard, especially in the final movement, was an open acknowledgment of death and at the same time a quiet celebration of the tranquillity connected to the process. I took this music as a metaphor for reassurance, confirming my own strong hunch that the dying of every living creature, the most natural of all experiences, has to be a peaceful experience. I rely on nature. The long passages on all the strings at the end, as close as music can come to expressing silence itself, I used to hear as Mahler's idea of leave-taking at its best. But always, I have heard this music as a solitary, private listener, thinking about death.

Now I hear it differently. I cannot listen to the last movement of the Mahler Ninth without the door-smashing intrusion of a huge new thought: death everywhere, the dying of everything, the end of humanity. The easy sadness expressed with such gentleness and delicacy by that repeated phrase on faded strings, over and over again, no longer comes to me as old, familiar news of the cycle of living and dying. All through the last notes my mind swarms with images of a world in which the thermonuclear bombs have begun to explode, in New York and San Francisco, in Moscow and Leningrad, in Paris, in Paris, in Paris. In Oxford and Cambridge, in Edinburgh. I cannot push away the thought of a cloud of radioactivity drifting along the Engadin, from the Moloja Pass to Ftan, killing off the part of the earth I love more than any other part.

I am old enough by this time to be used to the notion of dying, saddened by the glimpse when it has occurred but only transiently knocked down, able to regain my feet quickly at the thought of continuity, any day. I have acquired and held in affection until very recently another sideline of an idea which serves me well at dark times: the life of the earth is the same as the life of an organism: the great round being possesses a mind: the mind contains an infinite number of thoughts and memories: when I reach my time I may find myself still hanging around in some sort of midair, one of those small thoughts, drawn back into the memory of the earth: in that peculiar sense I will be alive.

Now all that has changed. I cannot think that way anymore. Not while those things are still in place, aimed everywhere, ready for launching.

This is a bad enough thing for the people in my generation. We can put up with it, I suppose, since we must. We are moving along anyway, like it or not. I can even set aside my private fancy about hanging around, in midair.

What I cannot imagine, what I cannot put up with, the thought that keeps grinding its way into my mind, making the Mahler into a hideous noise close to killing me, is what it would be like to be young. How do the young stand it? How can they keep their sanity? If I were very young, sixteen or seventeen years old, I think I would begin, perhaps very slowly and imperceptibly, to go crazy.

There is a short passage near the very end of the Mahler in which the almost vanishing violins, all engaged in a sustained backward glance, are edged aside for a few bars by the cellos. Those lower notes pick up fragments from the first movement, as though prepared to begin everything all over again, and then the cellos subside and disappear, like an exhalation. I used to hear this as a wonderful few seconds of encouragement: we'll be back, we're still here, keep going, keep going.

Now, with a pamphlet in front of me on a corner of my desk, published by the Congressional Office of Technology Assessment, entitled *MX Basing*, an analysis of all the alternative strategies for placement and protection of hundreds of these missiles, each capable of creating artificial suns to vaporize a hundred Hiroshimas, collectively capable of destroying the life of any continent, I cannot hear the same Mahler. Now, those cellos sound in my mind like the opening of all the hatches and the instant before ignition.

If I were sixteen or seventeen years old, I would not feel the cracking of my own brain, but I would know for sure that the whole world was coming unhinged. I can remember with some clarity what it was like to be sixteen. I had discovered the Brahms symphonies. I knew that there was something going on in the late Beethoven quartets that I would have to figure out, and I knew that there was plenty of time ahead for all the figuring I would ever have to do. I had never heard of Mahler. I was in no hurry. I was a college sophomore and had decided that Wallace Stevens and I possessed a comprehensive understanding of everything needed for a life. The years stretched away forever ahead, forever. My great-great grandfather had come from Wales, leaving his signature in the family Bible on the same page that carried, a century later, my father's signature. It never crossed my mind to wonder about the twenty-first century; it was just there, given, somewhere in the sure distance.

The man on television, Sunday midday, middle-aged and solid, nice-looking chap, all the facts at his fingertips, more dependable looking than most high-school principals, is talking about civilian defense, his responsibility in Washington. It can make an enormous difference, he is saying. Instead of the outright death of eighty million American citizens in twenty minutes, he says, we can, by careful planning and practice, get that number down to only forty million, maybe even twenty. The thing to do, he says, is to evacuate the cities quickly and have everyone get under shelter in the countryside. That way we can recover, and meanwhile we will have retaliated, incinerating all of Soviet society, he says. What about radioactive fallout? he is asked. Well, he says. Anyway, he says, if the Russians know they can only destroy forty million of us instead of eighty million, this will deter them. Of course, he adds, they have the capacity to kill all two hundred and twenty million of us if they were to try real hard, but they know we can do the same to them. If the figure is only forty million this will deter them, not worth the trouble, not worth the risk. Eighty million would be another matter, we should guard ourselves against losing that many all at once, he says.

If I were sixteen or seventeen years old and had to listen to that, or read things like that, I would want to give up listening and reading. I would begin thinking up new kinds of sounds, different from any music heard before, and I would be twisting and turning to rid myself of human language.